Zero-Sustainment Aircraft for the U.S. Air Force

A Workshop Summary

Gregory Eyring, Rapporteur

Committee on Zero-Sustainment Aircraft for the U.S. Air Force: A Workshop

Air Force Studies Board

Division on Engineering and Physical Sciences

NATIONAL RESEARCH COUNCIL
OF THE NATIONAL ACADEMIES

THE NATIONAL ACADEMIES PRESS
Washington, D.C.
www.nap.edu

THE NATIONAL ACADEMIES PRESS 500 Fifth Street, NW Washington, DC 20001

NOTICE: The project that is the subject of this report was approved by the Governing Board of the National Research Council, whose members are drawn from the councils of the National Academy of Sciences, the National Academy of Engineering, and the Institute of Medicine.

This is a report of work supported by Grant FA9550-12-1-0413 between the U.S. Air Force and the National Academy of Sciences. Any opinions, findings, or conclusions expressed in this publication are those of the author(s) and do not necessarily reflect the view of the organizations or agencies that provided support for the project.

International Standard Book Number 13: 978-0-309-27261-2
International Standard Book Number 10: 0-309-27261-0

Limited copies of this report are available from:

Air Force Studies Board
National Research Council
500 Fifth Street, NW
Washington, DC 20001
(202) 334-3111

Additional copies are available from:

The National Academies Press
500 Fifth Street, NW
Keck 360
Washington, DC 20001
(800) 624-6242 or (202) 334-3313
http://www.nap.edu

Printed in the United States of America

THE NATIONAL ACADEMIES

Advisers to the Nation on Science, Engineering, and Medicine

The **National Academy of Sciences** is a private, nonprofit, self-perpetuating society of distinguished scholars engaged in scientific and engineering research, dedicated to the furtherance of science and technology and to their use for the general welfare. Upon the authority of the charter granted to it by the Congress in 1863, the Academy has a mandate that requires it to advise the federal government on scientific and technical matters. Dr. Ralph J. Cicerone is president of the National Academy of Sciences.

The **National Academy of Engineering** was established in 1964, under the charter of the National Academy of Sciences, as a parallel organization of outstanding engineers. It is autonomous in its administration and in the selection of its members, sharing with the National Academy of Sciences the responsibility for advising the federal government. The National Academy of Engineering also sponsors engineering programs aimed at meeting national needs, encourages education and research, and recognizes the superior achievements of engineers. Dr. Charles M. Vest is president of the National Academy of Engineering.

The **Institute of Medicine** was established in 1970 by the National Academy of Sciences to secure the services of eminent members of appropriate professions in the examination of policy matters pertaining to the health of the public. The Institute acts under the responsibility given to the National Academy of Sciences by its congressional charter to be an adviser to the federal government and, upon its own initiative, to identify issues of medical care, research, and education. Dr. Harvey V. Fineberg is president of the Institute of Medicine.

The **National Research Council** was organized by the National Academy of Sciences in 1916 to associate the broad community of science and technology with the Academy's purposes of furthering knowledge and advising the federal government. Functioning in accordance with general policies determined by the Academy, the Council has become the principal operating agency of both the National Academy of Sciences and the National Academy of Engineering in providing services to the government, the public, and the scientific and engineering communities. The Council is administered jointly by both Academies and the Institute of Medicine. Dr. Ralph J. Cicerone and Dr. Charles M. Vest are chair and vice chair, respectively, of the National Research Council.

www.national-academies.org

COMMITTEE ON ZERO-SUSTAINMENT AIRCRAFT FOR THE U.S. AIR FORCE: A WORKSHOP

CLAUDE M. BOLTON, JR., Defense Acquisition University
CLAUDE V. CHRISTIANSON, National Defense University
THOM J. HODGSON, North Carolina State University
RONALD MUTZELBURG, Boeing Phantom Works (retired)
LYLE H. SCHWARTZ, University of Maryland at College Park
RAYMOND VALEIKA, Delta Airlines (retired)

Staff

CARTER W. FORD, Program Officer
SARAH M. CAPOTE, Research Associate
MARGUERITE E. SCHNEIDER, Administrative Coordinator

AIR FORCE STUDIES BOARD

GREGORY S. MARTIN, GS Martin Consulting, *Chair*
BRIAN A. ARNOLD, Raytheon Company
CLAUDE M. BOLTON, Defense Acquisition University
STEVEN R.J. BRUECK, University of New Mexico
THOMAS J. BURNS, Science Applications International Corporation
FRANK CAPPUCCIO, Cappuccio and Associates, LLC
DONALD C. FRASER, Charles Stark Draper Laboratory (retired)
MICHAEL J. GIANELLI, The Boeing Company (retired)
DANIEL HASTINGS, Massachusetts Institute of Technology
PAUL G. KAMINSKI, Technovation, Inc.
ROBERT LATIFF, R. Latiff Associates
NANCY G. LEVESON, Massachusetts Institute of Technology
MARK J. LEWIS, IDA Science and Technology Policy Institute
LESTER L. LYLES, The Lyles Group
MATT L. MLEZIVA, Wildwood Strategic Concepts
C. KUMAR N. PATEL, Pranalytica, Inc.
GERALD F. PERRYMAN, JR., Independent Consultant
RICHARD V. REYNOLDS, The VanFleet Group, LLC
J. DANIEL STEWART, University of Tennessee
REBECCA WINSTON, Winston Strategic Management Consulting

Staff

TERRY J. JAGGERS, Director
JESSICA R. BROKENBURR, Financial Assistant
SARAH M. CAPOTE, Research Associate
GREGORY EYRING, Senior Program Officer
CARTER W. FORD, Program Officer
CHRIS JONES, Financial Manager
MARGUERITE E. SCHNEIDER, Administrative Coordinator
DANIEL E.J. TALMAGE, JR., Program Officer

Preface

The Air Force recognizes that sustainment of legacy weapon systems is a strategic issue for the United States. To assist the Air Force in addressing this issue, the Air Force Studies Board of the National Research Council drafted terms of reference (TOR) in April 2012 for a short workshop to bring together Department of Defense organizations and industry to highlight current sustainment practices that the Air Force might leverage to reduce maintenance and sustainment costs in the near term. The National Research Council approved the TOR in July 2012. The 3-day workshop was then held on December 4-6, 2012, at the National Academy of Sciences Building in Washington, D.C.[1]

The committee is grateful for the support of the Air Force champion of this workshop, Lt Gen Judith Fedder, Deputy Chief of Staff for Logistics, Installations, and Mission Support, Headquarters Air Force. Lt Gen Fedder articulated a set of clear desired outcomes for the workshop prior to the workshop and in person at the workshop. In addition, the committee thanks the many expert speakers and guests who contributed to this activity. Finally, the committee's role was limited to planning the workshop, and the workshop summary has been prepared by the workshop rapporteur as a factual summary of what occurred at the workshop.

Claude M. Bolton, Jr., *Chair*
Committee on Zero-Sustainment Aircraft for the U.S. Air Force: A Workshop

---

[1]This is the second in a series of workshops conducted by the Air Force Studies Board at the request of the U.S. Air Force. It follows an earlier workshop titled "Energy Reduction at U.S. Air Force Facilities Using Industrial Processes," held on November 5-7, 2012.

Acknowledgment of Reviewers

This report has been reviewed in draft form by individuals chosen for their diverse perspectives and technical expertise, in accordance with procedures approved by the National Research Council's (NRC's) Report Review Committee. The purpose of this independent review is to provide candid and critical comments that will assist the institution in making its published report as sound as possible and to ensure that the report meets institutional standards for objectivity, evidence, and responsiveness to the study charge. The review comments and draft manuscript remain confidential to protect the integrity of the deliberative process. We wish to thank the following individuals for their review of this report:

Claude V. Christianson, National Defense University,
Nancy G. Leveson, Massachusetts Institute of Technology,
Eli Reshotko, Case Western Reserve University, and
Raymond Valeika, Delta Airlines (retired).

Although the reviewers listed above have provided many constructive comments and suggestions, they were not asked to endorse the views presented at the workshop, nor did they see the final draft of the workshop summary before its release. The review of this workshop summary was overseen by Wesley L. Harris, Massachusetts Institute of Technology. Appointed by the NRC, he was responsible for making certain that an independent examination of this workshop summary was carried out in accordance with institutional procedures and that all review comments were carefully considered. Responsibility for the final content of this summary rests entirely with the author and the institution.

Contents

APPENDIXES

Acronyms

ACAT	acquisition category
AFLCMC	Air Force Life Cycle Management Center
AFMC	Air Force Materiel Command
AFRL	Air Force Research Laboratory
AFSAC	Air Force Security Assistance Center
AFMC	Army Force Materiel Command
AFSC	Air Force Sustainment Center
AIM	automotive information module
ALC	Air Logistics Complex
AMC	Air Mobility Command
APU	auxiliary power unit
BCA	business case analysis
CFLI	core function lead integrator
CLS	contractor logistics support
DAU	Defense Acquisition University
DLA	Defense Logistics Agency
DLR	depot-level repair
DoD	Department of Defense
E2E	end to end
ERP	enterprise resource planning
FY	fiscal year
ISR	intelligence, surveillance, and reconnaissance
KPP	key performance parameter
KSA	key system attribute
LCC	life cycle cost
LHA	Logistics Health Assessment

LO	low observable
MAJCOM	major command
NAVAIR	naval aviation
NRC	National Research Council
OCO	overseas contingency operations
OEM	original equipment manufacturer
P&W	Pratt and Whitney
PEO	Program Executive Office
PM	program manager
PSI	product support integrator
PSM	program support manager
R&D	research and development
S&T	science and technology
TOR	terms of reference
VEMSO	Vehicle and Equipment Management Support Office
WSS	weapon system sustainment

Overview

Overall Air Force weapon system sustainment (WSS) costs are growing at more than 4 percent per year, while budgets have remained essentially flat. The cost growth is due partly to aging of the aircraft fleet and partly to the cost of supporting higher-performance aircraft and new capabilities provided by more complex and sophisticated systems, such as the latest intelligence, surveillance, and reconnaissance platforms. Furthermore, the expectation for the foreseeable future is that sustainment budgets are likely to decrease so that the gap between budgets and sustainment needs will likely continue to grow wider. Most observers accept that the Air Force will have to adopt new approaches to WSS if it is going to address this problem and remain capable of carrying out its missions.

In this context, the original intent of this 3-day workshop was to focus on ways that science and technology (S&T) could help the Air Force reduce sustainment costs. However, as the workshop evolved, the discussions focused more and more on Air Force leadership, management authority, and culture as the more critical factors that need to change in order to solve sustainment problems. Many participants who spoke at the workshop commented that while S&T investments could certainly help—particularly if applied in the early stages ("to the left") of the product life cycle—what is also important is adopting a transformational management approach—down to the shop level—that defines the user-driven goals of the enterprise, empowers people to achieve them, and holds them accountable. Several workshop participants urged Air Force leaders to start the process now, even though it will take years to percolate down through the entire organization. These sustainment concerns are not new and have been studied extensively, including in recent reports from the National Research Council's Air Force Studies Board and the Air Force Scientific Advisory Board.[1,2]

---

[1]NRC. 2011. *Examination of the U.S. Air Force's Aircraft Sustainment Needs in the Future and Its Strategy to Meet These Needs*. Washington, D.C.: The National Academies Press. Available at http://www.nap.edu/catalog.php?record_id=13177.

[2]Air Force Scientific Advisory Board. 2011. *Sustaining Air Force Aging Aircraft into the 21st Century*. Available at http://www.dtic.mil/cgi-bin/GetTRDoc?Location=U2&doc=GetTRDoc.pdf&AD=ADA562696. Last accessed December 27, 2012.

POSSIBLE ACTION ITEMS FOR AIR FORCE CONSIDERATION

Box O-1 contains potential actions that could be implemented within 6 months, which were suggested by various workshop participants to enable the Air Force to begin to address its ever-increasing sustainment costs.

BOX O-1 Possible Action Items Suggested by One Workshop Participant for Air Force Consideration

A. Initiate a sustainment pilot project, championed by the Air Force chief of staff and led by the Air Force Materiel Command commander, partnering with another Major Command, using the Navy's NAVAIR sustainment program as a template to:[1]

1. Manage Air Force weapon system sustainment (WSS) as an integrated enterprise that cuts across program boundaries.
2. Define a user-driven outcome the Air Force intends to achieve for the selected system, and describe the high-level supporting metrics that will be used to measure progress toward this outcome.
3. Decide who is the single individual or office responsible for managing Air Force WSS costs.
4. Define a simple, standard tool to use for a system's sustainment business case analysis that includes visibility over all actual sustainment costs incurred.
5. Establish or enhance transparency of total sustainment costs across the system's life cycle as well as across all Air Force sustainment and operational organizations.

B. Utilize the CORONA conference[2] mechanism to reach agreement among 4-star process owners as to the outcome metric to be used for the pilot program.

[1]The transformation of the Naval Aviation Enterprise went well beyond solely the application of "Lean" principles and into wide-ranging organizational and cultural changes.

[2]CORONA conferences are held three times a year allowing the secretary of the Air Force, the chief of staff, and senior Air Force military leaders to come together for open discussions on issues relevant to the Air Force's future.

1

Introduction

Each year, the Air Force faces a growing gap between the sustainment needs of its weapons and its annual sustainment budget. Overall weapon system sustainment (WSS) costs are growing at more than 4 percent per year while budgets have remained essentially flat. The cost growth is due in part to aging of an aircraft fleet (the average age is 23 years) that is suffering from increasing corrosion and fatigue cracking, with the attendant difficulty of finding replacement parts that are no longer in production and software written in languages that are no longer used. Costs are also rising due to the need to support higher-performance aircraft and new capabilities provided by more complex and sophisticated systems, such as the latest intelligence, surveillance, and reconnaissance (ISR) platforms. Furthermore, the expectation for the foreseeable future is that sustainment budgets are likely to decrease, so that the gap between budgets and sustainment needs will likely continue to grow wider. One workshop presenter suggested that the cost of ownership may be more threatening to aircraft than the enemy. Several participants noted that the Air Force will have to adopt new approaches to WSS if it is going to address this problem and remain capable of carrying out its missions.

These sustainment concerns are not new. The issue has been extensively studied, including in recent studies by the Air Force Studies Board of the National Research Council (NRC) and the Air Force Scientific Advisory Board.[1,2] There is recognition that part of the answer lies in bringing consideration of a weapon system's entire life cycle into the early planning and design phases of the weapon's acquisition process. Design choices such as materials, fasteners, and so on can have a big impact on maintenance costs, and principles, such as modular design and quick disconnects between modules, can aid in reducing disassembly and replacement costs. Numerous recommendations have also been made that address the way the Air Force organizes and manages its sustainment efforts—with many suggesting that the Air Force should

---

[1] NRC. 2011. *Examination of the U.S. Air Force's Aircraft Sustainment Needs in the Future and Its Strategy to Meet These Needs*. Washington, D.C.: The National Academies Press. Available at http://www.nap.edu/catalog.php?record_id=13177.

[2] Air Force Scientific Advisory Board. 2011. *Sustaining Air Force Aging Aircraft into the 21st Century*. Available at http://www.dtic.mil/cgi-bin/GetTRDoc?Location=U2&doc=GetTRDoc.pdf&AD=ADA562696. Last accessed December 27, 2012.

manage sustainment as an integrated enterprise, rather than as a series of parallel efforts for the various weapons programs.

The Air Force has begun to take a more integrated view of sustainment through, for example, consolidating sustainment responsibilities within the Air Force Materiel Command (AFMC) and organizing itself around eight core functions, each with an individual designated as a core function lead integrator. It remains to be seen whether these organizational changes will help to break down barriers to a more integrated approach to sustainment, although several workshop participants commented that there were opportunities for positive change.

WORKSHOP TERMS OF REFERENCE

The terms of reference for this workshop are given in Box 1-1.

> **BOX 1-1 Terms of Reference**
>
> An ad hoc committee will plan and convene one 3-day public workshop to (1) discuss how science and technology can reduce aircraft sustainment costs in the Air Force and (2) review costs in maintenance, upgrades, and aging aircraft in the Air Force.
>
> The committee will develop the agenda for the workshop, select and invite speakers and discussants, and moderate the discussions.
>
> In organizing the workshop, the committee might also consider additional topics close to and in line with those mentioned above. The workshop will use a mix of individual presentations, panels, breakout discussions, and question-and-answer sessions to develop an understanding of the relevant issues. Key stakeholders will be identified and invited to participate. One individually authored workshop summary document will be prepared by a designated rapporteur.

WORKSHOP STRUCTURE, SCOPE, AND APPROACH

The 3-day workshop, which occurred on December 5-7, 2012, in Washington, D.C., consisted of a series of presentations by invited speakers (biosketches of the committee members are provided in Appendix A; the workshop agenda is provided in Appendix B), with each presentation followed by general discussion. Broadly, the first day was devoted to presentations by Air Force and Department of Defense academic personnel; the second day to presentations on experiences within the other services and industry contractors; and the final half-day to discussion among all participants.

The original intent of this workshop was to focus on how the Air Force's science and technology (S&T) dollars should be spent to reduce sustainment costs, as suggested by Task 1 in the terms of reference (Box 1-1). Indeed, the workshop participants did hear from representatives of the Air Force Research Laboratory on its S&T investments and from several presenters in the military services and in industry regarding cases in which technology insertion had saved sustainment dollars.

Chapter 2 of this report provides a summary of the presentations delivered at this workshop and of the discussion that followed. Chapter 3 summarizes the discussion that occurred on the last day, organized into the following five general topic areas: (1) leadership and management; (2) mission statement and metrics; (3) setting budget priorities and funding; (4) relationships with the contractor community; and (5) culture issues and training.

2

Presentations and Comments

The workshop participants heard a series of presentations on sustainment challenges and initiatives within the military services and in private sector companies representing the aircraft industry. Abstracts of these presentations are provided in Appendix D. A brief summary of the main points of the presentations and the ensuing discussion is given next, in chronological order of presentation.

TUESDAY, DECEMBER 4, 2012

Lt Gen Judith Fedder, Deputy Chief of Staff for Logistics, Installations and Mission Support, Headquarters, U.S. Air Force

Lt Gen Judith Fedder, Deputy Chief of Staff for Logistics, Installations and Mission Support, Headquarters, U.S. Air Force, is responsible for weapon system sustainment (WSS) for the Air Force. Lt Gen Fedder noted that the elements of life-cycle WSS costs are spares/consumables, manpower, sustaining support, and depot maintenance.[1] These costs can be put in four categories: depot purchased equipment maintenance (33 percent), contractor logistics support (CLS; 61 percent), technical orders (1 percent), and sustaining engineering (5 percent). WSS baseline funding with supplementals has been about 80 percent of current requirements, but future funding is expected to fall further and further behind requirements.

CLS costs are driving overall WSS cost growth; CLS is growing at 7.9 percent per year, compared with 4.3 percent per year for WSS generally. The recent emphasis on intelligence, surveillance, and reconnaissance (ISR) aircraft is one driver, since these are contractor-maintained. In general, sustainment of newer weapon systems tends to be more contractor-heavy. Speed in acquisition also tends to work against organic sustainment.[2] Lt Gen Fedder concluded by listing some initiatives to enhance sustainment cost management:

---

[1]Lt Gen Judith Fedder, Deputy Chief of Staff for Logistics, Installations and Mission Support, Headquarters, U.S. Air Force, "Air Force Studies Board Workshop: Zero-Sustainment Aircraft," presentation to the workshop on December 5, 2012.

[2]For the purposes of this workshop summary "organic sustainment" is defined in the following way: "Organic logistics infrastructure refers to U.S. government entities (principally DoD organizations) such as inventory control

- Strategy guidance/tools/governance to life-cycle management community for building affordable/effective product support strategies;
- Enterprise-level initiatives such as NextGen CLS; and
- Individual program initiatives, including sustainment partnerships.[3]

Following the presentation, one observer noted that the F-22 was supposed to need only one-half of the maintenance required by the F-15 and asked, what happened? According to Lt Gen Fedder, the F-22 was envisioned to not need low observable maintenance specialists, just a "mech-tech" provided by the OEM. This turned out not to be the case. In addition, the Air Force added an egress system and other specialties and had to buy back the manpower to support these. There is a higher confidence that the F-35 will not need so many specialties, and this should result in large savings compared with the F-22.

Several participants raised critiques of the way the Air Force estimates the total costs of sustainment. One suggested a better breakdown of costs into five components that captures 95 percent of the costs: (1) maintenance, repair, and overhaul; (2) training; (3) personnel; (4) energy; and (5) modifications and upgrades. Because costs are incurred in so many different places, controlled by different authorities, rules, and colors of money, intelligent investment decisions cannot be made. A participant noted that the Air Force needs to view sustainment from a fleet perspective. The first question needs to be, What is the best way to deliver support? This participant offered as an example a similar situation in the Army. The M1 tank upgrade and fuel accounts were in different places, so when the question arose as to whether to put diesel in the tank, it was not possible to consider this from a business perspective. How can these components be brought together under one person?

Another participant remarked that data systems are not available that would serve as the basis for making smarter sustainment decisions. Knowing what people are actually doing is the key to cutting costs. Focusing on budgets is not the same as focusing on costs. Budgets reflect expenditures, not costs. The Air Force is budget-driven, but costs are more important. He noted that with respect to the growth of CLS costs, the Air Force dug itself a hole when it failed to purchase technical data at acquisition, which would have allowed the option to bring sustainment in-house. At the same time, the Air Force has lost technical expertise due to increase reliance on contractors, so the Air Force does not have access to what is driving CLS costs higher. On the organic side, this participant noted that another piece of the problem is that the supply chain and the depot are separate. "Which drives the bus?" The airlines have the same problem of knowing actual cost, but operationally the supply chain is under control of a

---

points, maintenance depots, distribution warehouses, and transportation facilities. Like the garden variety organic farmer who uses only natural or self-produced products, organic infrastructure sustainment uses the government's ability to support a product's mechanical and structural demands, such as those seen by the C-17, over the course of its life." Albert Barnes and Capt Lewis Johnson, U.S. Air Force, Going Organic: The C-17 Depot Maintenance Activation Working Group, Defense *AT&L Magazine,* September-October, 2010. Available at http://www.dau.mil/pubscats/ATL%20Docs/Sep-Oct10/Barnes%20sept-oct10.pdf. Accessed January 15, 2013.

[3]Lt Gen Judith Fedder, Deputy Chief of Staff for Logistics, Installations and Mission Support, Headquarters U.S. Air Force, "Air Force Studies Board Workshop: Zero-Sustainment Aircraft," presentation to the workshop on December 5, 2012.

single manager/user. The same participant stated that a recent analysis suggested that a typical time required to get an ordered part in the Air Force is 17 days.[4]

Another comment related to the organization of the Air Force, to the effect that the Air Force looks at problems through discrete views, or soda straws (e.g., AF/A1-A4), and cannot combine these views to link smart sustainment choices with acquisition decisions. The Air Force cannot get its arms around costs and, therefore, has no ability to influence the acquisition side. The Air Force cannot win the battle looking through these lenses. In Figure 2-1, for example, what are the options for cutting? Where can the Air Force cut to minimize risk? The answers are not clear. Yet, another participant asked rhetorically how much it costs to fly the F-15, all elements of cost together, fully burdened. The answer is that the Air Force does not know. The cost per flying hour may be known, but these data are not inclusive. The goal is to bring all of the communities together to agree on a metric. And if the Air Force has the data, does it have a decision mechanism to set priorities?

Another participant asked if the organic sustainment piece is going down as CLS goes up. He stressed that if the Air Force does outsourcing correctly, it needs to do it in a way that reduces internal staff levels to reduce costs. Related to that, he asked if the Air Force has a depot strategy. He answered his own question by commenting that Air Force should look at ISR and decide what skill set to keep in house, but that this strategy has not been developed. According to the participant, with platforms rushed to deployment during wartime, such as certain ISR platforms or the Army mine-resistant, ambush protected vehicle, DoD needs to plan up front for how these platforms will be brought into the regular sustainment system after the war is over.

A final topic of discussion related to the need to bring life cycle considerations into the acquisition process from the beginning. Lt Gen Fedder stated that the Air Force is well aware of the need to do better in this regard. One participant observed that the Air Force does not have the decision tools to make trade-offs early in the acquisition process. In the last 25 years, the division between acquisition and sustainment has gotten worse. In his view, program managers (PMs) should be fleet life-cycle managers and work closely with logistics experts. The Air Force currently does not have the authority, tools, or visibility to affect other parts of the product life cycle. This issue was recognized in the 1970s, but 40 years later has not been resolved.

Katherine Stevens, Director, Materials and Manufacturing Directorate and Capability Lead for Agile Combat Support, Air Force Research Laboratory

Katherine Stevens, Director, Materials and Manufacturing Directorate and Capability Lead for Agile Combat Support, Air Force Research Laboratory (AFRL), gave an overview of the AFRL's role in science and technology (S&T) for sustainment in the near term, mid-term, and far term. She also cited a number of examples of successful development and transition of sustainment technologies. Near-term activities address challenges in the maintenance of the current fleet, such as improved nondestructive inspection tools and expertise in support of the

[4]For a commercial airline of approximately 350 aircraft, cutting 1 day in the work-in-progress cycle—that is, from removal to repair to return to service—could save $7 million in inventory investment. This could be as simple as returning a failed part to a repair facility faster, cutting repair time, among other factors.

Air Force corrosion enterprise. Mid-term goals are to support "condition-based maintenance" in part by improving life prediction tools and sustainability models. Far-term goals include reducing sustainment costs by integration of data, models, and simulations throughout the life cycle.[5] When a physical aircraft arrives for maintenance, a digital model of the aircraft—specific to that tail number, including deviations from the nominal design—is intended to be delivered as well. The model is planned to continually reflect the current state of the actual aircraft.

Dr. Stevens believes the central reason behind increasing sustainment costs is the increasing age of the Air Force fleet; the average aircraft has been in service for 23 years. Problems associated with aging include fatigue cracking, corrosion, parts unavailability, material degradation, and wear. AFRL develops solutions to technology readiness level 6 and then hands them off to depots or program offices. Asked about where AFRL's requests for S&T support come from, Dr. Stevens indicated that signals come from both major commands (MAJCOMs) and program offices on warfighter needs. Projects from the program offices tend to focus on specific problems. The core function master plans provide more general guidance, and the MAJCOMs provide sustainment technology plans. Dr. Stevens indicated that it is challenging to decide how to invest in S&T for long-term (25-year) payback. The fidelity of cost-benefit analysis is inconsistent. The Air Force does not have a base process to track return on investment. One participant noted that in fairness, the commercial aircraft industry does not do this very well either.

An important issue raised was that of whether there are appropriate incentives for reducing sustainment costs. Dr. Stevens noted that there is no incentive for a program office to, for instance, cut in half the replacement times for landing gear on the F-22 unless it is given a requirement. Sometimes there are contractor incentives for improvements, but these are rare. It would also be possible to incentivize organic depots to find innovative ways to reduce sustainment costs, but this is not currently part of the culture. She observed that contract structure is the key to providing incentives—for example, a fixed-cost contract with a 50/50 cost share of any documented savings. Cost-plus contracts do not provide these incentives. Dr. Stevens believes that current contracting practices hold back full industry participation in reducing sustainment costs.

One participant noted that it had been about 2 years since the publication of the NRC's report on Air Force sustainment. It appeared to him from the presentation that big changes were happening at AFRL, but he questioned whether the sustainment budget at AFRL was still in the range of 5-8 percent of the overall budget. Dr. Stevens responded that in the past, AFRL focused exclusively on S&T research, but now sustainment is part of the mission. While it was true that many AFRL staff feel that it is more exciting to work on the cutting edge of technology, there are also AFRL people who are passionate about keeping the Air Force flying. This is a continuing challenge in AFRL culture.

In response to a question about whether there is duplication of effort between AFRL engineers and life-cycle engineers in the program offices, one participant felt that a bigger issue was retaining good life-cycle engineers in the program offices because they can receive higher

[5] Katherine Stevens, Director, Materials and Manufacturing Directorate and Capability Lead for Agile Combat Support, Air Force Research Laboratory, "USAF Science and Technology for Sustainment," presentation to the workshop on December 4, 2012.

salaries in the private sector. A different participant noted that a key element of reducing maintenance costs is to be able to identify the part that needs to be replaced right away, rather than having to follow a fault isolation tree. This requires good data on current component performance. Comparisons with the commercial airlines are not fair because the airlines have much younger fleets, and most aircraft are still in production. This is a big advantage for maintenance and parts replacement.

A final comment relating to personnel was that, in years past, development planning tools existed with which the Air Force defined capability gaps in the next 20 years, and experts ran models and simulations on all weapon systems, which became the basis for the program objective memorandum submission to Congress. This all went away when Congress zeroed out funding. The funding has now been restored, but the experts have now retired.

Steven Brown, Professor, Defense Acquisition University

Steven Brown, a professor at Defense Acquisition University and a former Air Force crash investigator, began by observing that flight safety is a passion in the Air Force. In his view, the Air Force could solve its sustainment problems if it had the same passion. The key is to institutionalize progressive change. His presentation was organized around five key areas of sustainment: cost, performance, management, contracting, and training. In each area, he described the current state of affairs, recent innovations, and, finally, suggestions for institutionalizing these changes.[6]

He estimated that about 65 percent of total ownership cost of an aircraft is operating and support (O&S) cost. It is very difficult to get a handle on total cost because the Air Force does not have data on life-cycle cost (LCC). In the acquisition process, O&S cost is a key system attribute (KSA) and must be estimated to check a box, but there is no hardcore requirement pertaining to it (it is not a key performance parameter, or KPP), and it is not tracked. Dr. Brown remarked that acquisition and LCC are not connected in the Air Force. They are managed by two completely separate organizations at the system program offices (SPOs). He suggested that improved analysis and tracking of LCC could be made a requirement.

Some participants related two anecdotes about poor decisions that were made because of a failure to consider life-cycle costs up front. In the F-22, the decision was made not to carry a ladder on board in order to save weight. This necessitated every airfield to have deployable ladders on hand to serve the F-22s, at much higher cost. In another example, the Army proposed that the future combat system carry bottled water, since the cost of delivering water to the vehicles was so large. However, the cost of delivery was borne by a different office, and it proved impossible to make a business case for the vehicles to carry the extra weight. Some participants noted that the lesson drawn is that program offices must have the tools available to make the correct decisions where these kinds of life-cycle trade-offs are involved.

One participant agreed that if money is important, the Air Force has to track it in order to make proper decisions, and the Air Force has to have appropriate metrics. A common metric is dollars per flying hour, but that does not account for capability. The Air Force needs to know

---

[6]Steven Brown, Professor, Defense Acquisition University, "Institutionalizing Low Sustainment Aircraft," presentation to the workshop on December 4, 2012.

what elements of cost are included in the estimates in order to compare with industry. In the airline industry, maintenance cost per available seat mile is the metric. Utilization is also key. In the airline industry, aircraft commonly fly 3,000-4,500 hours per year, much more than Air Force aircraft.

Dr. Brown suggested that the Air Force look outside for best practices to benchmark against. The Navy has been emphasizing "gate reviews" early in the acquisition process that have been demonstrated to reduce O&S cost uncertainty, and the Air Force could learn from this. Pratt and Whitney and others have developed useful cost models. Finally, other countries such as the United Kingdom and Canada realize lower sustainment costs because they have historically operated under tighter dollar constraints, and there are likely to be lessons to be learned there.

On the performance issue, Dr. Brown noted that system availability (measured as percent time available) is typically a KPP, while system reliability (measured in terms of mean time between failures) is often a lower level KSA. This was the case with the Joint Strike Fighter (JSF), where "mission reliability" (availability) was the KPP. He believes that this decision is one reason that the growth in system reliability of some versions of the JSF has lagged behind what was planned. One participant questioned what the availability metric really means. It is not clear in the field what the "fudge factors" are. At the strategic level, this is not a concern. Is the aircraft flying or not? But at the tactical level, it is likely that different metrics are being used for example, for aircraft moving through the depots. To institutionalize improvements to sustainment performance, Dr. Brown recommends establishing system requirements for reliability and maintainability.

On the management issue, DoD Instruction 5000 describes a systems engineering process for weapons programs in which the PM is also the "life-cycle systems manager," with no dedicated logistician in the loop. However, a new key leadership position of product support manager (PSM) has now been established for all acquisition category I and II (ACAT I and II) programs. This life-cycle logistician, who works directly for the PM, is OSD-certified at level 3 and is also supposed to be acquisition-certified. To institutionalize this advance, Dr. Brown recommended filling the PSM positions with highly qualified life-cycle logisticians and benchmarking sustainability readiness levels against maturity models in the other services, industry, and allied initiatives.

On the contracting issue, Dr. Brown noted the trend away from the PM being involved in the details of repairs, parts, and engineering toward a focus on performance-based logistics—availability, reliability, and mission effectiveness. The details of product support are increasingly being managed by product support integrators (PSIs) who work for the PSM. The PSIs are given contract incentives such as fees and 50/50 gain sharing as rewards for saving sustainment dollars. This strategy has been shown to work well for the F-117 and the F/A-18. The Defense Logistics Agency (DLA) has also initiated pilot programs that combine existing performance-based contracts for sustainment of common items (e.g., auxiliary power units (APUs), helicopter engines) managed by the different services into a single DoD contract. This enterprise approach to sustainment contracting is estimated to save about 20 percent on costs.

On the training issue, DoD is estimated to have nearly 17,000 life-cycle logisticians—some in the program offices, some in the supply chain, and some at the three Air Logistics

Complexes (ALCs)—94 percent of whom are civilians.[7,8] Defense Acquisition University (DAU) offers three levels of logistics certification and in 2013 will offer a new course in "business acumen" that will help acquisition personnel develop techniques for negotiating better business deals. In 2014, a senior seminar will be offered to current and selected PSMs that highlights the keys to PSM and PM success. A final comment following this talk was that DoD does not have the tools to measure return on investment or life-cycle cost, and instead of chief executive officers, chief financial officers, and quarterly reports, there is a political process that makes it hard to run sustainment in a businesslike manner.

James Yankel, Technical Director, Directorate of Logistics, Air Force Materiel Command

James Yankel, Technical Director, Directorate of Logistics, Air Force Materiel Command (AFMC), provided a perspective on sustaining aging aircraft from AFMC. Average growth of total aircraft sustainment costs is 6.5 percent per year, driven by increasing failure rates and maintenance man-hours associated with an aging fleet.[9] Funding is falling further and further behind WSS requirements. The biggest cost growth area is CLS. In the 1960s, the majority of sustainment was organic. Today, the majority is commercial contracts. In the future, the plan is to make sustainment more of a public-private partnership. Mr. Yankel described recent initiatives aimed at reducing sustainment costs and approaching sustainment more from an enterprise point of view. He explained that AFMC recently reorganized itself to achieve efficiencies by consolidating 12 centers down to the following 5:

- Air Force Research Laboratory (AFRL);
- Air Force Life Cycle Management Center (AFLCMC);
- Air Force Sustainment Center (AFSC);
- Air Force Test Center; and
- Air Force Nuclear Weapons Center.

In addition, processes for sustainment at AFMC are increasingly taking a life cycle and command-wide approach. These processes are undergoing performance reviews that

---

[7]The number of DoD personnel who are responsible for life-cycle logistics, which is defined as "developing, fielding, and improving system sustainment," is large; however, this number is relatively small when compared to the number of people (~600,000) who are responsible for broader DoD logistics (Steven Brown, Defense Acquisition University, "Institutionalizing Low Sustainment Aircraft," presentation to the workshop on December 4, 2012).

[8]"The U.S. Air Force (USAF) currently has three Air Logistics Centers [Complexes] (ALCs), operating under the Air Force Materiel Command (AFMC), which provide acquisition, modification, and maintenance support for the Air Force aircraft fleets, end items, commodity parts, and some missile systems. The ALCs are complex, multi-faceted organizations. They provide support to the Air Force and other components of the Department of Defense (DoD) on numerous product lines." Excerpted from NRC, *Examination of the U.S. Air Force's Aircraft Sustainment Needs in the Future and Its Strategy to Meet These Needs,* Washington, D.C.: The National Academies Press, 2011.

[9]James Yankel, Technical Director, Directorate of Logistics, Air Force Materiel Command, "Sustaining Aging Aircraft," presentation to the workshop on December 4, 2012.

emphasize standardized reporting across all weapon systems, requirements cost drivers, and technology insertion needs and opportunities.

Following the presentation, some participants raised questions about whether what is driving the cost of sustaining each weapon system is truly understood by the Air Force. The issue is not whether sustainment is accomplished by CLS or depot-level repair (DLR), but rather what is the best cost solution for the Air Force. Other participants pointed out that discussions of rising costs should consider operational tempo as a factor to determine if costs are reasonable. One participant opined that flight hours should be the common denominator, not average age. In the commercial airline industry, aircraft availability (or its opposite, downtime) based on the need for structural repairs is the cost driver, not repair cost.

A final topic of discussion was the AFMC reorganization and whether it is likely to produce the desired efficiencies. Some participants noted that this was an open question so soon after the reorganization, with the reorganization still in the "sausage-making" phase. Several observers saw a potential conflict between the responsibilities of the AFLCMC and the AFSC. There was sentiment expressed by some participants that there should be a strong cooperative relationship between the two centers and that the two complementary responsibilities should be clearly articulated and institutionalized to produce a common approach.

Joann Berrett, Director, Aerospace Sustainment Directorate, Air Force Sustainment Center

Joan Berrett, Director, Aerospace Sustainment Directorate, AFSC, began by stating that there is a high level of effort directed toward maintaining good communication between AFSC and AFLCMC. All elements of AFSC organic costs are being investigated, although she was not sure when the results would be available. The three major Air Logistics Complexes (ALCs)—Oklahoma City ALC; Warner Robins ALC; and Ogden ALC—are organized under AFSC.[10] She noted that the number of dollars needed to satisfy readiness requirements are projected to increase while the budget for WSS is expected to be flat from FY2012 to FY2018. AFSC's aim is to produce higher efficiencies through better processes and a higher level of integration. She cited some success stories in which average work flow days had been reduced through implementation of high-velocity manufacturing principles—for example, improved knowledge of an aircraft's condition before it arrives at the depot, and "gated production," in which all relevant repairs on an aircraft are completed before it can move on to the next gate.

Following this talk, there was a discussion of the Defense Logistics Agency's role. While some of DLA's successes were acknowledged by some participants—especially in the area of contracting, one participant claimed that it takes a very long time—17 days on average—to get an ordered part. Another observed that one reason is that the DLA does not own all the assets it needs—e.g., transportation—and that the Army's record is even worse. A final comment on DLA was that there is no integrated supply chain planning across the enterprise. DLA needs to be more agile and remove compartmentalization. Ms. Berrett was asked whether the "not invented here" syndrome still prevailed at the three ALCs. She answered that Lt Gen Litchfield

[10]Joan Berrett, Director, Aerospace Sustainment Directorate, Air Force Sustainment Center, "Driving to Cost Effective Readiness," presentation to the workshop on December 4, 2012.

had initiated a big push to standardize work processes and metrics at the ALCs, from the mechanic on the shop floor up to the senior managers.

Mike Jennings, Deputy Director of Logistics (Acting), Air Force Life Cycle Management Center

Mike Jennings, Deputy Director of Logistics (Acting), Air Force Life Cycle Management Center, described the makeup of AFLCMC, which focuses on acquisition and includes the program executive offices (PEOs). With the new five-center construct described above, Lt Gen C.D. Moore has more visibility across the enterprise, although no one person is empowered to execute an enterprise sustainment strategy. A Life Cycle Management Working Group was established that reviewed more than 160 processes and identified needs in the following six key areas owned by AFLCMC:

- *Develop human capital.* There is currently no standard training for specialties. There is now only a very limited capability to identify needed competencies and what AFLCMC now has, although there is an implementation plan for competency of the new PSM positions and other logisticians.
- *Developmental planning.* AFLCMC already possesses an "executive" role. This will be critical to having the ability to influence product support.
- *Product support business case analyses (BCAs).* Standard processes have been defined but, according to Mr. Jennings, still need oversight.
- *Repair sources.* Cross-program efficiencies need to be identified (for example, common radios on different platforms).
- *Centralized asset management and WSS prioritization.* Requirements standardization and justification are needed, as is a standard methodology for looking across platforms at engineering requirements.
- *Logistics health assessments (LHAs).* There is currently no ability to roll up sustainment data to the enterprise level; assessment and reporting are not lined up. A tool for milestone B to C exists (engineering and manufacturing development).[11]

The common thread among all of these needs is that there are many functional communities—engineering, budget, and so on—that need to work together. Processes need to be integrated—e.g., BCAs with depot source of repair. In Mr. Jennings' view, Lt Gen Moore needs to have power over the product support enterprise, that is, influence over PEOs.

Mr. Jennings stated that the overall objective of AFLCMC is to provide affordable, effective product support. Metric number one is system availability, although the definition of availability is still under discussion. Affordability is judged by looking at the acquisition program baseline and comparing to actual expenditures. This will help in estimating future expenditures. Metric number two is LHAs, although he stressed that the Air Force must have the necessary data.

---

[11]Mike Jennings, Deputy Director of Logistics (Acting), Air Force Life Cycle Management Center, "Product Support Responsibilities and Cost Reduction Initiatives," presentation to the workshop on December 4, 2012.

Mr. Jennings was asked who measures actual costs. He responded that currently the issue is not resolved and needs to be tied to firm data and updated when major events occur. This is still part of the "sausage making" at AFMC. A participant stressed that it is very important to check the BCA against actual costs. This need not be complicated, and it should not be necessary to clear the results with everyone unless there is a big deviation from forecasts. He noted that the fundamental challenge is to define an Air Force life-cycle cost model that will be used to plan and track costs. To be truly useful, it must be tied to budgets.

Final Thoughts—Day 1

Following this presentation, the chair asked for final thoughts from individual participants based on the presentations and discussions from the first day. The following individual views, which do not necessarily reflect the consensus of the workshop participants and speakers as a group, were expressed:

- Based on the positive discussion, it appears that the Air Force has the right leaders and the right people on the ground to address the sustainment problem. With regard to the aging fleet, the Air Force has not done a good job of going to the people who have worked these issues for three decades to learn lessons of how they have succeeded. In particular, the Air Force Security Assistance Center (AFSAC) supports more aircraft than are in the Air Force inventory. AFSAC's inventory includes WWII vintage aircraft all the way to the most modern aircraft. AFSAC has developed the expertise to mitigate a number of sustainment issues that the operational Air Force is facing. When asked whether or not AFSAC expertise had been tapped, the answer on three different occasions throughout the first day was no.
- The emphasis on standardization across the organization is positive. Implementing the "Lean" approach is not a short-term process—it is likely to take 8 years. The structural changes at AFMC are encouraging, but the Air Force lacks the right data and information to do a top-notch sustainment job. The challenge is all the more difficult with rotations of personnel.
- The problems caused by colors of money are discouraging and getting worse. They are also self-inflicted. This would be a good subject for further study.
- The discussion is reminiscent of the 1960s. The Air Force needs to track costs, and the life-cycle management effort should focus on a few low-hanging fruit. It needs a near-term success.
- The Air Force needs to define a life-cycle cost model for sustainment and disseminate it.
- There have been enormous changes in the past 2 years, and it appears that the recommendations of the NRC's sustainment report have been heard. However, continuity of leadership remains a problem. How can the Air Force institutionalize change?

WEDNESDAY, DECEMBER 5, 2012

Maj Mark Blumke, Deputy Chief, Mx Systems and Integration Branch, Directorate of Logistics, Air Mobility Command

Maj Mark Blumke, Deputy Chief, Mx Systems and Integration Branch, Directorate of Logistics, Air Mobility Command (AMC), reviewed AMC initiatives to reduce maintenance and sustainment costs.[12] He listed the top cross-cutting drivers, in order of effects on operational availability or non-mission-capable hours (these do not reflect direct labor hours, but instead are measured on a 24-hour clock). Engines appear second on the list, but in fact are the top cost driver. He stated that engine initiatives at AMC are driven by fuel-efficiency programs but also have the secondary effect of reducing sustainment costs. He presented several BCAs for engine upgrades in which the avoided maintenance costs were projected to yield savings on the same order as the fuel savings. However, one participant noted that no actual cost data was presented and stressed the importance of following up with actual data to compare with projected data. Furthermore, specific fuel consumption needs to be measured in a test cell, since a lot of extraneous factors come into play when trying to measure fuel consumption on a flying airplane. Another participant noted that constraints on the modifications budget were preventing cost-effective engine projects from going forward.

Another AMC effort that is intended to increase fuel efficiency and reduce maintenance is to reduce the use of APUs on aircraft when they are on the ground, since external generators are five times more efficient. One participant observed that putting clocks on APUs—and making them easy to read—would help in this effort. Maj Blumke went on to discuss structure-related initiatives at AMC, the most important of which is corrosion. He cited the 2010 report *Impact of Corrosion on Cost and Availability to DoD* that estimated that Air Force aviation and missile corrosion consumes 35 percent of maintenance costs, roughly $6.5 billion.[13] Many AMC aircraft are high on the list. Initiatives include funding development of corrosion prediction and growth models (particularly corrosion under paint) and sensors to monitor corrosion initiation and propagation. Corrosion is a serious issue, and AMC takes it seriously. AMC stands ready to work with other entities within the Air Force and DoD to advance corrosion prevention/control capabilities.

Many observers commented that if corrosion truly accounts for 35 percent of costs, it should be the object of a major cross-cutting S&T effort. Several noted that many studies by AFRL, the NRC, and others have already been done, and there was a funded program managed by the Joint Logistics Command to address corrosion across the Air Force. A participant questioned, What happened? One participant familiar with the effort said a lot of coatings were developed but that the initiative appeared to lose steam after awhile as ideas got used up.

---

[12]Maj Mark Blumke, Deputy Chief, Mx Systems and Integration Branch, Directorate of Logistics, Air Mobility Command, "AMC Initiatives to Reduce Maintenance and Sustainment Cost Drivers," presentation to the workshop on December 5, 2012.

[13]Logistics Management Institute, 2010, *Impact of Corrosion on Cost and Availability to DoD.* Available at http://www.sae.org/events/dod/presentations/2012/impact_of_corrosion_on_cost_and_availability_to_dod.pdf. Accessed February 22, 2013.

A final comment on the presentation made three points. First, in the commercial airline industry, every time a panel is opened (e.g., during an inspection) the mechanic uses a basic corrosion protection spray. This has been in place for 15 years and should be routine. Does the Air Force have a comparable procedure? Second, if the Air Force is really focusing on cost to the enterprise, it should have a cost-related metric such as man-hours, as opposed to non-mission-capable hours. Third, the engine initiatives described appeared to be related to occasions on which major modifications were undertaken, but there was less emphasis on the smaller things that can be done. There are not that many modifications performed on old aircraft.

SMSgt Kevin Mead, Air Force Element Vehicle and Equipment Management Support Office

SMSgt Mead's organization is chartered to execute enterprise vehicle fleet management and sustainment for the Air Force. It is customer-focused at the base level and has authority to act without going through any MAJCOM. He described several initiatives that have saved man-hours and sustainment costs, including validating fleet requirements and rightsizing, utilizing a standard algorithm for budget forecasting, and introducing fleet health metric with an algorithm to measure the health and effective ages of the 97,000-vehicle fleet. The Vehicle and Equipment Management Support Office (VEMSO) has developed a fleet management decision support system with utilization criteria and validation processes. In addition, there is a program called automotive information module (AIM-2), to be completed by the end of FY2013, in which an installed module collects data on the vehicle usage characteristics, fuel type, etc., and transmits the information wirelessly to a worldwide database that is searchable by customers. This is enabling a condition-based maintenance approach.

The initiatives described in this presentation were praised in comments made by many participants who thought the fleet management perspective and the enterprise approach as refreshing. SMSgt Mead was asked if any other Air Force programs had come in to see the tools he was using. He responded that he had been visited by the construction and civil engineering communities and that tools are available on the community website. VEMSO reports directly to headquarters of the Air Force, which is very unusual.

One participant stressed that automated reporting of problems on ground vehicles is far in advance of the situation with aircraft. Aircraft need to be made "smart" so that they can tell engineers of any problems. Another observer said that the bigger problem is that everyone looks differently at metrics. The key metric is dollars for the enterprise. The best investment might be an anti-corrosion spray, not an engine modification. The PSMs recommend where to spend the next dollar in a particular program; the Air Force must create an environment where these recommendations can be made enterprise-wide. An example of the problem is the inability to change configuration and buy a new engine that would be better in all respects than doing a modification. In addition, the constraints associated with using specific funding areas are extremely burdensome. These funding rules are counterproductive to reducing sustainment and maintenance costs. A participant argued that the Air Force must empower fleet managers with the ability to expedite moving money from one area to another if the business case can be made.

Lt Col Brian Godfrey, Chief, Airborne Branch (A4CA), Air Combat Command Headquarters

Lt Col Brian Godfrey, Chief, Airborne Branch (A4CA), Air Combat Command Headquarters, gave an overview of sustainment from an Air Combat Command (ACC) perspective. The biggest cost drivers are structural cracks/corrosion, wing repairs/replacements, wiring faults, and low observable (LO) maintenance. These problems are exacerbated by fewer maintainers available, increasing operational tempo, and fleet aging. Sustainment cost successes include better fleet management—e.g., balancing the time spent by F-15s and F-16s in high- and low-corrosion environments— and shifting F-22 maintenance from contractor to organic. In his view, contractor logistical support is not a panacea. In the future, he believes S&T can be harnessed to save on maintenance in several ways: common/standardized testers; leveraging information technology (e.g., having mobile maintenance applications on a tablet carried by the mechanic to be able to troubleshoot, order parts, and close jobs at the aircraft); alternative fuels; and "cold spray" of metal onto surfaces to repair corrosion, cracks, and holes. Lt Col Godfrey was enthusiastic about the young, tech-savvy workforce at the depots and felt that they were an asset that could help implement new maintenance technologies.

One participant asked why the Air Force is still having problems with LO maintenance. Participants asked, Will the Air Force continue to have problems with the F-35? Did the Air Force learn the appropriate lessons from experience with the F-117 and the B-2 structures? The speaker remarked that one problem was that the Air Force did not have enough people then, and this is still true. Lt Col Godfrey was also asked whether he was proposing a standard test program. He responded that there had been such a program years ago that was a big failure. The lesson was to be careful of global information technology solutions. Another participant responded that the Air Force needs to define common data architectures across the enterprise, not common testers. The discussion then centered around enterprise resource planning (ERP) systems that are designed to improve business processes but are not designed specifically to enhance decision making. The data ERPs generate do enable better decisions, but the ERP itself is not a good decision-support tool. One participant noted that what would aid significantly in decision making would be an Air Force "app store" that would offer easy-to-use applications that would run off of the ERP, much like applications for today's smart phones.

Another comment was that the Air Force should make a business case for putting a "black box" on legacy aircraft, with sensors that tell where and when problems arise. The problem is how to fund this if it competes with funding for aircraft performance. A response by one participant was that there is money dedicated to the sustainment community that does not compete with performance—you just have to figure out where the money is. If corrosion is the most important problem, the Air Force needs to figure out how to put money against it. Who in the Air Force could make this decision? Program offices put out requirements for brochures on depot-level maintenance for that weapon system, but the Air Force does not have that at the corporate level. Centralized asset management just orchestrates the process program by program, but the Air Force is missing opportunities unless the Air Force looks across the board and implements changes broadly. Lt Col Godfrey offered that the core function lead integrator (CFLI) construct is one way to do this, but it is core function by core function, not enterprise wide. There is a disconnect between programs and the enterprise. In the commercial world, there is an FAA-approved manual that prescribes how to remove paint, and this is used for all products.

One participant remarked that LCMC should prescribe standards by which programs are managed—standard data elements that allow the Air Force to do what it needs to do. Another commented that LCMC should establish a task force on corrosion to synergize efforts; it might require 5 years to figure out how to do it, but the Air Force would know what it is spending on it. The Air Force can use information technology to streamline and customize tasks in such a way that users will inform the enterprise—the opposite of the traditional military chain of command. One observer noted that PMs are generally not in the life-cycle cost business. In the Army, through engine designs that took advantage of modules and quick disconnects, it would be possible to change out a Humvee engine in 2 hours instead of the current 32 hours, but who would pay for it? DoD needs to create a global supply chain; this is not just an Air Force problem. A final comment was that data mining is a powerful tool for figuring out where to position spare parts or services most efficiently. It was found, for example, that by mining data of people googling "cold and flu," this was the best predictor of where flu vaccines should be sent.

BG Edward Dorman III, Director for Logistics Operations, Readiness, Force Integration and Strategy, Office of the Deputy Chief of Staff, G-4

BG Edward Dorman III, Director for Logistics Operations, Readiness, Force Integration and Strategy, Office of the Deputy Chief of Staff of the Army, G-4, addressed two topics: (1) operational energy consumption and (2) condition-based reset. Soldiers are using more fuel and battery energy than ever before, and energy drives operational capability—maneuver, awareness, communication, etc. The fully burdened cost of fuel ranges from $3.95/gallon to more than $56/gallon in Afghanistan. Batteries carried by soldiers are an important weight issue. The Army G4, chief of staff, and the secretary of the Army are leading a variety of initiatives to promote energy conservation and energy efficiency, including soldier-worn integrated power enhancement systems, wheeled vehicle systems to recharge batteries, and engine upgrades to improve efficiency. However, making soldiers more energy-aware requires training and changing the culture of senior non-commissioned officers and soldiers; the officers must not be the only ones to see the benefits.

Condition-based reset is an effort to get back to soldier maintenance in the field—toward organic unit-based as opposed to institution-based sustainment of equipment to save money. The examples given tended to focus on ground equipment rather than aviation. Condition-based reset has secondary impacts on reducing operational energy by reducing "tooth to tail" the entire logistics snake. One participant noted that sustainment solutions are not always materiel solutions; he liked the emphasis on gathering more information and the "condition-based" concept. Another participant stressed the importance of institutionalizing this approach, so that it would survive personnel rotations, and getting buy-in, down to the foxholes.

Joe Guenther, Vice President and General Manager, Evandale Turbofan and Turbojet Engines, General Electric Aviation

Joe Guenther, Vice President and General Manager, Evandale Turbofan and Turbojet Engines, General Electric Aviation, explained that factors driving engine sustainment costs are the operating environment, which is not under the warfighter control, the aircraft mission (for example, the decision to use the 4-engine B-2 bomber for close air support in Afghanistan), and thrust de-rate—only using full thrust when the airplane is full. Factors that can strongly affect maintenance costs are engine health monitoring, inspection practices, and regular engine washes. It is important to have the analytical capability to monitor engines that should come off the airplane today rather than fail in 3 weeks. If GE has a performance-based logistics contract, it monitors trends in engine health with electronic controls wirelessly. In fact, Mr. Guenther showed examples where sustainment managed by the original equipment manufacturer (OEM) resulted in lower cost per shop visit and longer time on wing. Engine fleet management is also important. The best engines are not needed for an easy route. One sees a lot of "tired" engines in the Air Force. As engines age, they become more expensive to restore.

New turbine engine technology will reduce fuel consumption and increase thrust-to-weight ratio. To take advantage of emerging capabilities, Mr. Guenther recommended continuing to expand the component improvement program, funding upgrades to the legacy fleet, re-engineering, and continuing investment in new technology/materiel. The discussion following this presentation was wide-ranging. Several observers related problems with the interpretation of existing regulations—e.g., what could be funded by 3400 (operations and maintenance) money versus 3010 (modifications) money. One participant opined that the challenge is to determine how sustainment translates into things that matter to the warfighters/operators. The platforms meet the mission but cost more dollars than they should. The participant commented that declining budgets should be viewed as opportunity to effect change. Another observer commented on the effects of the new CFLI construct on sustainment. Each quarter, "red" and "yellow" (non-combat ready) aircraft are reported up to the CFLI. One core function, agile combat support, represents 70 percent of the inventory, but the current process only addresses segments of the problem.

Raymond Valeika, Retired Senior Vice President for Technical Operations, Delta Airlines

Raymond Valeika, Retired Senior Vice President for Technical Operations, Delta Airlines, drew lessons from his management experience in the commercial airline industry for the Air Force. He explained that effective sustainment depends on transparent information, based on the following:

- *Manufacturers' data.* The relationship between the military and the OEMs is not as close as the relationship between the commercial airlines and the OEMs. There needs to be constant communication.

- *Operational data.* Having proper metrics is important. At Delta, the goal was to have no more than 12 aircraft undergoing unscheduled maintenance and 18 in scheduled maintenance out of a fleet of 600. These numbers were tracked daily.
- *Cost data.* It is critical to understand all elements of cost, including inventory, labor, aircraft downtime, overhead, and staff. Application of good sustainment management principles, such as focusing comprehensively on one aircraft at a time, saves money. Cost data also allow one to make trade-offs; e.g., one-engine taxi can save fuel but can hasten the failure of the other engine due to lack of warm-up time. *If saving money is the goal, it is critical to manage costs rather than manage budgets.*
- *Performance data.* One must understand and validate error rates. How many aircraft are out of service, and what does that mean? These data help determine training needs—typically 1 week per employee per year at Delta. Sick leave and absentee rates are key labor metrics that are tracked because they reflect employee attitudes toward work.

Mr. Valeika stated that transparent information drives organizational effectiveness. The organization must be integrated in order to capture the gains made by either operations or sustainment personnel. There has to be a place where authority comes together to resolve conflicts. The Air Force has a matrix structure that does not provide measures across the boundaries. For example, the user of a part is not responsible for having the part delivered. Cost is looked at in stovepipes. Landing gear on the F-15s were taken from spares and put on operational aircraft, leaving a field full of aircraft with no feet. This was done to get the mission accomplished, but at what cost?

Following this presentation, one observer noted that Gen Wolfenbarger, head of AFMC, is seeking an enterprise viewpoint of all centers in order to be able to tell the chief of staff what is going on. The three ALCs were once commanded by two-star flag officers and saw one another as competitors. With the recent reorganization, they now all report to Lt Gen Bruce Litchfield, commander of AFSC, who now has more clout, budget, and access. In this observer's view, this is a big opportunity.

Mark Buongiorno, Director, Military Engine Aftermarket Business Development, Pratt and Whitney

Mark Buongiorno, Director, Military Engine Aftermarket Business Development, Pratt and Whitney (P&W), reviewed life cycle cost management practices at P&W. He noted that engine design largely determines sustainment requirements. In terms of the life cycle, he estimated that development is 11 percent of the cost, production 23 percent, and sustainment 66 percent. Cost drivers change over the life cycle, with depot-level repair being about 30 percent of 5-year recurring costs but nearly 75 percent of 50-year recurring costs. He stressed that active life-cycle management is required in all program phases and cited examples of cases where P&W's approach had saved the Air Force money. P&W works with the depots to determine when engines get overhauled. It presents metrics and cost drivers to the operators—a very rich data exchange. Finally, P&W has an investment program to continuously improve its products through technology insertion. Following this talk, one observer commented on the differences between military and commercial aircraft engine utilization. Military aircraft are flown much differently than commercial aircraft, leading to shorter engine life on the military side. Effective sustainment requires a serial-number-specific sustainment work program.

Final Thoughts—Day 2

Following this presentation, the chair asked for final thoughts from individual participants based on the presentations and discussions from the first day. The following individual views, which do not necessarily reflect the consensus of the workshop participants and speakers as a group, were expressed:

- There is a perceived deterioration in relationships between industry contractors and government acquisition personnel compared to the past. This theme recurred several times during the workshop.
- Depots have developed clever workarounds when they encounter problems with the system. These lessons should be captured.
- The Air Force does not have a simple answer to the question of what its mission is. The Navy, on the other hand, knows. The Air Force corporate structure is now organized around core functions, but the CFLI process has not been internalized yet.
- Color of money is a major issue.
- A process is lacking for converting cost-saving opportunities into a business case. The resources may be available if the business case is made.
- The Air Force does not have a meaningful cost tool to translate sustainment issues into the budget framework. It does not have to be complex and should begin with "small c" cost.

THURSDAY, DECEMBER 6, 2012

VADM Walter Massenburg (USN, Ret.), Senior Director, Mission Assurance Business Execution, Raytheon Company

VADM Walter Massenburg (USN, Ret.), Senior Director, Mission Assurance Business Execution, Raytheon Company, recounted a situation in naval aviation in the 1999-2001 period in which aviation depot-level repairable (AVDLR) costs were rising by double digits each year, and spare aircraft were being cannibalized to support deployed aircraft. Only 30 percent of the fleet was flyable. Appalled by this situation, the chief of naval operations (ADM Clark at that time) instituted an enterprise vision of naval aviation on a cross-functional, cross-organizational behavior model aligned to the "greater good." The model used was a commercial one, with a chief executive officer, a chief operating officer, and a chief financial officer managing the aircraft fleet according to Lean management principles.

In the beginning, the Navy had no clue as to what was driving their costs. Strong egos were trying to protect individual programs. There was a culture oriented toward consumption; e.g., the metric being used on carriers was the number of arrested landings in a given time period. It took 4 years to break through the stovepipes and get rid of impediments, and they began to look at readiness differently. There was a single process owner (keeper of the metric). The metric adopted was "aviation units ready for tasking based on missions completed." This overall metric has a number of supporting metrics involving inventory, reliability, cycle time reduction, and total cost (all dollars/all financial stovepipes). Consistent with Lean principles, the goal is to achieve the metrics and no more. It is important to decide how much is enough; going beyond is not necessarily good and may be costly in both financial and personnel terms. The metric must be aligned with end-user value. One must (1) understand the outcome to be achieved, (2) define processes to achieve the outcome, and (3) reorganize only to the extent it affects the outcome.

VADM Massenburg cited a number of examples where this enterprise approach resulted in better fleet readiness at lower cost. Managers had to subordinate personal priorities for the greater good. Money left over at the end of the fiscal year was returned to the enterprise rather than spent. NAVAIR became the only organization recapitalizing its force, to the tune of $4 billion per year. This represented a "life spiral" rather than a death spiral. Much of the latter part of the presentation was devoted to encouraging productivity in the workforce. People are the source of capability to perform the mission and to achieve program success. It is the role of leadership to get the incentives right and inspire people. The culture should emphasize subordination to the metric, not collaboration per se. Work needs to be driven by demand pull, and worker talents need to be matched to the tasks. By emphasizing productivity, staff costs can be reduced without threatening the delivery of end products. VADM Massenburg concluded with a list of best practices and behaviors:

- Identify domains and assign single process owners.
- Assemble the right enterprise teams and gain commitment.
- Operate in support of a single fleet-driven metric (what the enterprise values).
  —Get agreement on scope, outputs, and linked metrics;
  —Make data transparent to promote trust and monitor performance;
  —Share knowledge on issues and key problems affecting the domain;
  —Recognize, nurture, and support technical authority; and
  —Identify entitlements (what's needed, when, and how much and no more).
- Agree on the desired output (e.g., readiness over cost), with focus on the trade space involving current and future readiness.
- Operate with discipline, governance, and a regular (timely) "drumbeat."
- Baseline every dollar, all the people, all the stuff, and all the capability within the domain, with assigned accountability for outcomes.
- Establish entitlements. Continually measure gaps to entitlement.
- Remove barriers to productivity.[14]

[14]Two books were recommended for further discussion of these principles: Gary Connors' *Lean Manufacturing for the Small Shop* (Society of Manufacturing, 2001) and Joel Levitt's *Lean Maintenance* (Industrial Press, 2008).

3

Wrap-Up Discussion

The final day was devoted primarily to general discussion and an attempt to distill the main points that had been presented by those who spoke at the workshop. The discussion involved the following topic areas: (1) leadership and management; (2) mission statement and metrics; (3) setting budget priorities and funding; (4) relationships with the contractor community; and (5) culture issues and training.

LEADERSHIP AND MANAGEMENT

Some workshop participants noted that the Air Force has recognized that the trend of year-over-year increases in sustainment costs cannot continue, especially with the prospect of budget reductions throughout DoD as the United States attempts to get its fiscal house in order. One participant asserted that unless the approach to sustainment changes, this will lead to a "death spiral" for the Air Force. However, another participant noted that the discipline of the anticipated budget cuts could be viewed as an opportunity to make fundamental changes in the Air Force's approach to sustainment.

Several questioned, however, whether there was leadership buy-in at the highest levels of the Air Force. Many observed that enterprise-level thinking does not occur in the Air Force. Some participants noted that policies must be articulated by the Air Force chief of staff and be internalized down to the wing level.

Echoing the conclusions of a previous study, several workshop participants noted that there is no single person responsible for sustainment throughout the Air Force.[1] Rather, sustainment decisions are made within the individual program "stovepipes," with no one having visibility across programs. One example cited was the issue of corrosion of parts and structures, which has been estimated to be responsible for up to 30 percent of sustainment costs across the Air Force. Each program wrestles with its own corrosion problems individually, whereas such a large, common problem would be better addressed holistically at the enterprise

---

[1]NRC. 2011. *Examination of the U.S. Air Force's Aircraft Sustainment Needs in the Future and Its Strategy to Meet These Needs*. Washington, D.C.: The National Academies Press. Available at http://www.nap.edu/catalog.php?record_id=13177.

level. In the commercial airline industry, for example, there are policies in place that address the minimization of corrosion across all types of aircraft in the fleet.

The recent reorganization within AFMC with the consolidation of 12 product centers down to 5 was seen by some who spoke at the workshop as an opportunity for taking a more holistic approach to sustainment. The leader of the new AFLCMC, for example, deals with all program stovepipes and, in principle, has the ability to influence (although not to make) sustainment decisions across programs. Similarly, the leader of AFSC has control of the workflow at maintenance depots and test centers. Some participants pointed out that a challenge for AFMC is to determine what it is trying to achieve as an outcome and to get all of its components to work together to achieve that outcome. While the reorganization within AFMC was noted as a positive development by some workshop participants, it was not seen as a panacea. Indeed, reorganization per se was not seen as the solution to sustainment issues across the Air Force. Rather, a key point made by the some workshop participants involved the need to articulate at the highest levels the user-driven enterprise outcome desired and the need to give key individuals the authority and accountability for achieving that outcome.

Some workshop participants noted that the success achieved in NAVAIR, which was driven primarily by customer-focused metrics tied to fully burdened costs and through the application of "Lean" management principles, provides an "existence proof" and template for what can be accomplished in the Air Force. However, participants who commented did not minimize the challenges involved in implementing these management principles. They noted that the NAVAIR example was not implemented across the Navy as a whole, but only within a specific part. Accordingly, some participants expressed the view that implementation within the Air Force should start with a pilot or prototype project focusing on one weapon system (e.g., C-130 or F-15) and involving AFMC working with another MAJCOM, though there was disagreement about which one would make the best partner. Some workshop participants stated that AMC would be the most appropriate partner given its combat support mission, which is similar to that of AFMC. Others felt that Air Combat Command (ACC) would be the right choice, given the importance of getting the operators involved.

There was also discussion of who might champion such a pilot program. By analogy with the chief of naval operations' role in the NAVAIR example, some participants who commented felt that the champion should be the Air Force chief of staff. However, it was noted that there needs to be another individual authorized to be the "campaign manager" who would facilitate training and changing peoples' behavior. Some participants suggested that this person should be a four-star general, perhaps the leader of AFMC. The chief would give this individual the goal and associated metrics and hold him or her accountable for achieving them. It was pointed out that the transformation that took place at NAVAIR was difficult and took place over a period of at least 4 years. Thus, the continuity of leadership is a critical issue. A participant noted that regular rotations of command personnel make it imperative that changes be institutionalized so that they do not depend on individuals who may only be in the job for 18 months.

MISSION STATEMENT AND METRICS

A central tenet of Lean is that the entire organization needs to focus on what creates value for the end customer, and the use of resources for any other goal is a waste. Implicit in this formulation is that sufficient effort should be expended to preserve customer value, but no more. Continuing with the Lean theme, the overall goal needs to be supported by high-level metrics that are tied to the end user—in the Air Force case, the person who pulls the trigger. These metrics need not be complicated. One participant noted that in the commercial airline industry he focused every day on three metrics: (1) safety; (2) regulatory compliance; and (3) basic operational numbers, such as number of aircraft out of service. Throughout the workshop, examples were given of Air Force metrics that reflect consumption of resources (e.g., number of sorties or flight hours), with the implicit assumption that more consumption is better, rather than focusing on value for the mission or the end user. The focus is more often on output than outcome. In the NAVAIR example, the metric became "aircraft ready for tasking based on missions completed." This metric, which evolved over time, focused not just on the readiness of aircraft flying missions at a given time, but also on the number available for training and for future missions. It thus minimized the practice of cannibalizing inactive aircraft in order to keep deployed aircraft flying. Finally, several participants noted that metrics need to be established for comptrollers that go beyond dollars obligated.[2]

SETTING BUDGET PRIORITIES AND FUNDING

One consequence of the program-oriented, stovepiped structure of the Air Force is the inability to set budget priorities based on cost to the enterprise as a whole; i.e., to answer the question, If the Air Force has only one additional dollar to spend, where should it be spent? The Air Force has no way of understanding and tracking total cost. The canonical example discussed at the workshop was the issue of corrosion. If, indeed, corrosion accounts for 30 percent of the Air Force sustainment budget, as some studies have suggested, a strong case can be made that the next dollar should be spent on a program to address it. With respect to sustainment, there is no mechanism for looking at budgets across programs, and, indeed, AFMC does not have visibility into ACC's sustainment budget. During the workshop, it was noted that a number of budget metrics are being tracked because of legislative or other requirements, such as non-mission-capable hours and contractor logistics support, but it was pointed out that these are irrelevant to the question of how best to reduce total sustainment cost to the Air Force. Some participants expressed that a broader view of costs is required to encompass costs to accomplish the mission of the joint force and total sustainment cost.

One concern expressed by a workshop participant related to the fragmented approach to costs in the Air Force is the difficulty of translating legitimate needs into a convincing business case for funding. One participant expressed the need for a standard tool for doing this, stressing that it would be only a tool and need not be complicated. An important corollary is the need to track actual costs and compare them with projected costs to verify that the

[2]The transformation of the Naval Aviation Enterprise went well beyond solely the application of Lean principals and into wide ranging organizational and cultural changes.

projected gains are in fact realized. While several presenters showed projected cost savings associated with specific sustainment programs, there were little data presented on actual costs incurred.

Finally, some participants expressed the view that restrictions on how money can be spent from the various accounts associated with sustainment ("color of money") were a barrier to reducing total sustainment costs. For example, in the case of substituting a new part for an old part, if a new part has the same form, fit, and function as the old part, the substitute can be funded from the operations and maintenance account; if not, it is considered a modification, and must be funded from the acquisitions and modifications account. A case was cited in which an internal engine part that would improve performance was proposed to replace an older part. However, substitution of the new part could not be funded because it did not have exactly the same form, fit, and function as the old part and was, therefore, considered to be a modification, even though the outside profile of the engine remained the same.

RELATIONSHIPS WITH THE CONTRACTOR COMMUNITY

Although it was not a major theme of the workshop, a company representative raised a "red flag" regarding what he perceived to be deteriorating relationships between government managers and the contractor community and what this could mean to Air Force sustainment in the future. This participant noted that it is difficult to get a feel for the Air Force contracting process, due in part to the distributed authority. He stated that many companies having both commercial and government businesses are rethinking their involvement with the government because of all of the audits and red tape they have to deal with. He cited one case in which a government request for information had to be changed because small businesses could not compete. In the past, he asserted that communications between the Air Force and industry were much better than today. Government personnel used to be mentored on how to work with industry, but in his view this has all gone away.

This situation contrasts sharply with the strong relationships that exist between his company and its suppliers in the commercial world. All are working together to improve safety and compliance, and it is rarely necessary to negotiate contracts to improve an engine. There is much more freedom to act. As a result, this company much prefers working with other companies and is having difficulty staffing positions relating to military programs. There are no engineers waiting "pencils in hand" to contribute. In his view, the warfighters are getting less. While the evidence provided by this participant was anecdotal, the passion with which these views were expressed was notable.

CULTURE ISSUES AND TRAINING

According to one participant with experience in the NAVAIR Lean management experiment, the biggest barrier to changing Air Force culture and breaking through the stovepiped organization is likely to be the egos of the managers involved. The current incentives reward the best stovepipe, except during wartime. He argued that to adopt Lean management, the culture will have to change—subordination to the goal of the enterprise rather than pursuit of personal priorities. The objective should be to manage costs, not budgets. The "use it or lose it" mentality on budgets will have to be changed. It should be encouraged to return unspent money to the enterprise at the end of the fiscal year.

This participant asserted that Lean management also requires that all members of the organization understand the business and its objectives through training. In the Air Force case, this includes both warfighters and civilians. Various educational institutions were suggested as providers of this training, including the Air War College, although this would not be available to civilians. However, Lean management is not just to be pursued at the enterprise level. The same participant stressed that the goal should be to give every first line supervisor Lean management training. In his experience, Lean management works well on the shop floor, and in his business, it led to reductions in both overhead and paperwork. Another participant with a background in the Army testified that Lean management implementation tools and performance metrics are available and helped to facilitate Army policy updates and to rationalize supply regulations.

Appendix A

Biographical Sketches of Committee Members

Honorable Claude M. Bolton, Jr., *Chair*, became the executive-in-residence for the Defense Acquisition University (DAU) on January 3, 2008. Mr. Bolton's primary focus is assisting the DAU president achieve the congressional direction to recruit, retain, train and educate the Department of Defense (DoD) acquisition workforce. He is also a management consultant to defense and commercial companies and is a board member for several companies. Prior to becoming the DAU executive-in-residence, Mr. Bolton served as the assistant secretary of the army for acquisition, logistics and technology (ASAALT). As the ASA (ALT), he served as the army acquisition executive, the senior procurement executive, and the science advisor to the secretary. Mr. Bolton oversaw the Elimination of Chemical Weapons Program and had oversight and executive authority over the Project and Contracting Office charged with Iraq reconstruction. He was responsible for appointing, managing, and evaluating program executive officers as well as managing the Army Acquisition Corps and Army Acquisition Workforce. Mr. Bolton retired as a Major General in the U.S. Air Force following a highly decorated career. Some highlights of his Air Force service include serving as the commander of the Air Force Security Assistance Center, where he managed foreign military sales programs with totals exceeding $90 billion that supported more than 80 foreign countries; serving as a test pilot for the F-4, F-111, and F-16; program executive officer for the Air Force Fighter and Bomber programs; and the first program manager for the Advance Tactical Fighter Technologies program, which evolved into the F-22 System Program Office. An experienced command pilot flying more than 40 different aircraft including Army helicopters, during the Vietnam War he flew 232 combat missions, 40 over North Vietnam. Mr. Bolton served as commandant of the Defense Systems Management College and as inspector general and director of requirements at Air Force Materiel Command headquarters. Mr. Bolton's education includes a bachelor's degree in electrical engineering from the University of Nebraska, a master's degree in management from Troy State University; and an M.A. in national security and strategic studies from the Naval War College. In 2006, he was awarded a D.Sc. (Honoris Causa) from Cranfield University. In May, he was awarded an Honorary Doctor of Science degree from the University of Nebraska. Mr. Bolton is a member of the NRC's Air Force Studies Board and is a past member of the Committee on Evaluation of U.S. Air Force Preacquisition

Technology Development and Committee on Optimizing U.S. Air Force and Department of Defense Review of Air Force Acquisition Programs.

Claude V. Christianson is director of the Center for Joint and Strategic Logistics at National Defense University. Prior to this position he served as the Chief Executive Officer of Global Logistics Associates LLC, an Alexandria, VA-based, member-owned, limited liability company specializing in professional logistics and supply chain services. Mr. Christianson's military career culminated as the Director of Logistics, J4, on the Joint Staff. As the J4 he synchronized joint logistics support across all Services and DoD agencies in support of operations worldwide. As the Army's Deputy Chief of Staff for Logistics, G4, Mr. Christianson drove the fielding of a commercially sourced satellite network to the logistics domain, connecting logisticians across forward-deployed, austere environments. Mr. Christianson served as the Chief of Logistics, C4, Coalition Land Forces Command (CFLCC), during Operation Iraqi Freedom in Kuwait from 2002-2003, where he directed the planning and execution of logistics support for more than 240,000 ground forces and over more than 300,000 square miles. From 2000-2002, as the Deputy C4 (Logistics) for Combined Forces Command, U.S. Forces Korea Director of Logistics, J4, and Eighth U.S. Army Deputy Commanding General (Support), Mr. Christianson directed the planning and execution of logistics operations in support of all combined and joint forces in Korea. From 1998-2000, Mr. Christianson served as Deputy Commanding General, 21st Theater Support Command, European Theater Support Command in Germany where he coordinated the execution of European logistics support for Operation Joint Guardian (Kosovo). Mr. Christianson is a distinguished military graduate of the North Dakota State University Army ROTC program and holds a B.S. in industrial engineering.

Thom J. Hodgson is the James T. Ryan Distinguished University Professor, an Alumni Distinguished Research Professor, co-director of the Operations Research Program, and director of Graduate Programs of Engineering-On-Line at North Carolina State University (NCSU). He served as director of the Integrated Manufacturing Systems Engineering Institute at NCSU (1995-2011); director of the Division of Design and Manufacturing Systems at the National Science Foundation (1991-1993); head of the Industrial Engineering Department at NCSU (1983-1990); professor of Industrial and Systems Engineering at the University of Florida (1970-1983); operations research analyst at Ford Motor Company (1966-1970); and an officer in the U.S. Army (1961-1963). He is a member of the National Academy of Engineering and is a fellow of IIE and INFORMS and is the author or co-author of more than 80 journal articles and book chapters. Dr. Hodgson served as a member of the U.S. Army Science Board (1994-2000) and is a current member of the NRC's Committee on Energy Reduction at U.S. Air Force Facilities Using Industrial Processes: A Workshop.

Ronald Mutzelburg retired from the Boeing Company as the Washington, D.C. director for the Phantom Works and Advanced Systems. His organization managed the relationship with senior U.S. government technology and advanced systems customers, including the Defense Advanced Research Projects Agency, Office of the Director, Defense Research and Engineering, the Office of Naval Research, NASA (Aeronautics), as well as the Office of the Secretary of Defense, Joint Staff, and Military Service technology and long-range capability requirements offices. Prior to joining Boeing, Mr. Mutzelburg completed a 34-year government career within the DoD where he served in the following positions: deputy director for air warfare, Office of Strategic and Tactical Systems, Under Secretary of Defense for Acquisition, Technology and Logistics, where he was responsible for acquisition oversight for the B-1, B-2, C-17, F-22, F-18, Joint Strike Fighter, JSTARS, Unmanned Air Vehicles, several proprietary programs, and numerous air-to-air and air-to-ground weapons programs; assistant program director for the B-2, Aeronautical Systems Division (ASD), Air Force Systems Command; director of fighter propulsion, Propulsion Systems Program Office, ASD; director of Logistics, Propulsion Systems Program Office, ASD. From 1982 to 1983, Mr. Mutzelburg attended the National War College. From 1968-1982, he held numerous managerial and project officer assignments within Air Force Logistics Command. He has received numerous awards and much recognition over the years, including the DoD Distinguished Civilian Service Award (2001) and the Presidential Rank Award. He has an M.S. in industrial and systems engineering from Ohio State University and is a graduate of National War College. Mr. Mutzelburg was a member of the NRC's Committee on Evaluation of U.S. Air Force Preacquisition Technology Development.

Lyle H. Schwartz retired from government service in 2004 after 18 years as a member of the Senior Executive Service. In his last position, as director, Air Force Office of Scientific Research (AFOSR), he guided the management of the entire basic research investment for the Air Force. He led a staff of more than 200 scientists, engineers, and support staff in Arlington, Virginia, and two foreign technology offices in London and Tokyo. As director he was charged with maintaining the technological superiority of the Air Force. Prior to becoming AFOSR's director, Dr. Schwartz directed the AFOSR's Aerospace and Materials Sciences Directorate. He is known for contributions to phase transitions in iron alloys, applications of Mossbauer spectroscopy, x-ray and neutron diffraction, characterization of catalysts, and policy issues concerning materials science and engineering. He is a member of the National Academy of Engineering and has written more than 85 technical papers and is co-author of two textbooks in materials science and engineering. Dr. Schwartz is a past member of the NRC's Air Force Studies Board and was member of the NRC's Committee on Examination of the U.S. Air Force's Aircraft Sustainment Needs in the Future and Its Strategy to Meet Those Needs.

Raymond Valeika is an independent consultant advising major companies in aviation matters. He is an internationally recognized aviation operations executive with more than 40 years of experience managing large airline maintenance operations, equally comfortable in the United States and abroad dealing with regulators, manufacturers and employees. Mr. Valeika retired from Delta as senior vice president for technical operations where he directed a worldwide maintenance and engineering staff of more than 10,000 professionals, maintaining a fleet of nearly 600 aircraft. Through his leadership and focus on continuous improvement of the human processes in aviation maintenance, Delta Technical Operations consistently rated at the top of the industry for performance benchmarks in the areas of safety, quality, productivity, and reliability. He is currently is on the board of the Flight Safety Foundation as well as on the board of AerCap, Inc., and SRT. In addition, he was senior vice president of technical operations at Continental and vice president of maintenance and engineering at Pan AM. He graduated from St. Louis University with a degree in aeronautical engineering. Mr. Valeika has served on previous NRC studies sponsored by the Air Force, including the Committee on Examination of the U.S. Air Force's Aircraft Sustainment Needs in the Future and Its Strategy to Meet Those Needs and the Committee on Analysis of Air Force Engine Efficiency Improvement Options for Large Non-fighter Aircraft.

Appendix B

Workshop Agenda

OBJECTIVES

1. Address the current state of sustainment.
   a. What factors are responsible for driving your systems' sustainment costs? Where is our sustainment money going today?
   b. How do those cost factors change over time?
   c. Have you been successful in more effectively "managing" system sustainment costs?
2. Address the potential for science and technology to impact our future costs.
   a. What emerging technologies show promise to reduce or eliminate those sustainment cost drivers?
   b. How are you pursuing those opportunities? What "approaches" would you recommend to take advantage of emerging capabilities that might reduce our sustainment costs?

TUESDAY, DECEMBER 4, 2012

0900 Welcome and Introductions

- The Honorable Maj Gen (Ret.) Claude M. Bolton, Jr., Workshop Committee Chair

0930 Vision for the Workshop

- Lt Gen Judith Fedder, Deputy Chief of Staff for Logistics, Installations and Mission Support, Headquarters U.S. Air Force

1000 Break

1015 Science and Technology for Sustainment

- Dr. Katherine Stevens, Director, Materials and Manufacturing Directorate, Air Force Research Laboratory

1115 Institutionalizing "Low Sustainment" Aircraft

- Dr. Steven Brown, Professor, Defense Acquisition University

1215 Continue Discussions *with Lunch Available*

1315 Air Force Materiel Command Initiatives

- *Aging Aircraft Maintenance*—Mr. James Yankel, Technical Director, Directorate of Logistics, Air Force Materiel Command
- *Driving to Cost-Effective Readiness*—Ms. Joann Berrett, Director, Aerospace Sustainment Directorate, Air Force Sustainment Center

1515 Break

1530 Air Force Materiel Command Initiatives Continued

- *Product Support Responsibilities and Cost Reduction Initiatives*—Mr. Mike Jennings, Deputy Director of Logistics (Acting), Air Force Life Cycle Management Center

1630 Workshop Committee Feedback to Day 1 Presentations

- All

1700 Adjourn

WEDNESDAY, DECEMBER 5, 2012

0900 Air Mobility Command Initiatives

- *Air Force Element (AFELM) Vehicle and Equipment Management Support Office (VEMSO) Initiatives to Reduce Sustainment Costs*—Maj Mark Blumke, Deputy Chief, Mx Systems and Integration Branch, Directorate of Logistics, Air Mobility Command

1000 Break

1015 Air Combat Command Initiatives

- Lt Col Brian Godfrey, Chief, Airborne Branch (A4CA), HQ Air Combat Command

1115 Army Initiatives

- *Informed Sustainment*—BG Edward Dorman III, Director for Logistics Operations, Readiness, Force Integration and Strategy, Office of the Deputy Chief of Staff, G-4

1215 Continue Discussions *with Lunch Available*

1315 Industry Initiatives Continued

- *GE Initiatives*—Mr. Joe Guenther, Vice President and General Manager, Evandale Turbofan and Turbojet Engines, General Electric Aviation
- *Sustainment: Managing Consequences of Failure with Transparent Information*—Mr. Raymond Valeika, Retired Senior Vice President for Technical Operations, Delta Airlines
- *Pratt and Whitney Initiatives*
  —Mr. Randy LaMar, Fleet Operations Discipline Chief, Pratt and Whitney
  —Mr. Mark Buongiorno, Director, Military Engine Aftermarket Business Development, Pratt and Whitney

1615 Workshop Committee Feedback to Day 2 Presentations

- All

1700 Adjourn

THURSDAY, DECEMBER 6, 2012

0900 Industry Initiatives Continued

- *The Enterprise Approach: Why Now?*—VADM Walter Massenburg (USN, Ret.), Senior Director, Mission Assurance Business Execution, Raytheon Company

1000 Break

1015 General Discussion with Participants to Include Next Steps

- All

1200 Continue Discussions *with Lunch Available*

1300 Adjourn

Appendix C

Workshop Participants

COMMITTEE MEMBERS

Honorable (Maj Gen [Ret.]) Claude M. Bolton, Jr., *Chair*
LTG (Ret.) Claude V. Christianson
Thom J. Hodgson (NAE)*
Ronald Mutzelburg
Lyle H. Schwartz (NAE)*
Raymond Valeika

NATIONAL RESEARCH COUNCIL STAFF

Terry Jaggers, Air Force Studies Board Director
Carter Ford, Program Officer
Greg Eyring, Rapporteur
Sarah Capote, Research Associate
Marguerite Schneider, Administrative Coordinator

SPEAKERS

Joann Berrett, Director, Aerospace Sustainment Directorate, Air Force Sustainment Center
Steven Brown, Professor, Defense Acquisition University
Mark Buongiorno, Director, Military Engine Aftermarket Business Development, Pratt and Whitney
BG Edward Dorman III, Director for Logistics Operations, Readiness, Force Integration and Strategy, Office of the Deputy Chief of Staff, G-4
Lt Gen Judith Fedder, Deputy Chief of Staff for Logistics, Installations and Mission Support, Headquarters, U.S. Air Force
Lt Col Brian Godfrey, Chief, Airborne Branch (A4CA), HQ Air Combat Command

---

*Member of the National Academy of Engineering.

Joe Guenther, Vice President and General Manager, Evendale Turbofan and Turbojet Engines, General Electric Aviation
Randy LaMar, Chief of Engine Logistics Program Management, Pratt and Whitney
VADM Walter Massenburg (USN, Ret.), Senior Director, Mission Assurance Business Execution, Raytheon Company
Katherine Stevens, Director, Materials and Manufacturing Directorate, Air Force Research Laboratory
James Yankel, Technical Director, Directorate of Logistics, Air Force Materiel Command

GUESTS

Joseph Baker, Deputy Capability Lead, Agile Combat Support, Air Force Research Laboratory/RX
Maj Michael Dunlavy, Air Force Materials and Manufacturing PEM, U.S. Air Force
David Madden, Division Chief, Product Support Engineering Division, Air Force Life Cycle Management Center, Air Force Materiel Command
Gary Reese, Director, Strategic Planning, Washington Operations, General Electric Aviation
Col Joe Robinson, Air Force/A4L
Cathy Snyder, Air Force Life Cycle Management Center
John Turco, General Electric Aviation
Angie Tymofichuk, Director, Engineering and Technical Management, Air Force Sustainment Center/EN
Marc Whitt, Senior Policy Analyst SAF/AQR

Appendix D

Presentation Abstracts

Speaker: Lt Gen Judith Fedder, Deputy Chief of Staff for Logistics, Installations and Mission Support, Headquarters U.S. Air Force

Presentation Title: A4/7 Sustainment

Multiple factors influence life-cycle costs: spares/consumables, manpower, sustaining support, fuel, depot Mx, other costs. Weapon system sustainment (WSS) is a key measure of component repair/overhaul; requirements are outpacing baseline funding, even with retirements. Air Force acquisition and sustainment communities are working to decrease these costs. Understanding sustainment factors and trade-offs early in the acquisition cycle is needed. Analysis of mission requirements, operational costs, life-cycle costs, reliability and speed of repairs, and using predictive maintenance tools and training is also necessary. Weighing contractor versus organic sustainment options and considering resources required for both the short- and long-term should be considered in this workshop. We are committed to evolving our logistics core competencies to posture logistics resources for the next fight and deliver cost-effective logistics readiness. Under the near term strategic priorities, we are committed to understanding and inform the cost of logistics and re-establishing expertise within life-cycle logistics. Considering resources to best meet readiness requirements effectively and efficiently is paramount, outlining the need for focused efforts.

Speaker: Katherine Stevens, Director, Materials and Manufacturing Directorate Air Force Research Laboratory

Presentation Title: Science and Technology for Sustainment

As the U.S. Air Force fleet continues to age, the cost of sustaining the fleet will consume an ever larger share of the Air Force budget. The Air Force Research Laboratory (AFRL) executes approximately $2 billion per year on science and technology efforts for the Air Force. A portion of these funds is directed towards sustainment, with an emphasis on keeping the current fleet safe, improving aircraft availability rates, reducing life-cycle costs (LCC) and improving the sustainability of future weapon systems. AFRL is executing an investment plan with near-, mid-, and far-term technology goals. Recent technology successes have resolved issues negatively impacting mission capable rates and resulted in multi-billion dollar LCC avoidance. Results of recent studies conducted by the National Academy of Sciences and the Air Force Scientific Advisory Board on the future of Air Force sustainment and how S&T is responding to the recommendations will be presented.

Speaker: Steve Brown, C.P.L., Defense Acquisition University

Presentation Title: Institutionalizing Low Sustainment Aircraft

During his presentation titled "Institutionalizing Low Sustainment Aircraft," Professor Steve Brown discussed five issues critical to successfully reducing sustainment of military air vehicles with the Air Force Studies Board at the National Academy of Sciences on December 4, 2012.

1. Sustainment cost requirements, data analysis, and reviews during the system life cycle;
2. RAM (reliability, availability, and maintainability) performance requirements and funding;
3. Emerging life-cycle program management best practices, including program support manager key leadership position and benchmarking of services, industry, and allied initiatives;
4. Contracting approaches to grow implementation of performance based logistics strategies; and
5. Enhanced Department of Defense (DoD) life-cycle workforce training, including new Defense Acquisition University courses focusing on business acumen and senior seminar for product support managers.

After highlighting current law and DoD policy related to each topic, examples of how lower aircraft sustainment costs have been achieved were summarized and considerations to institutionalize low-sustainment aircraft throughout the military service were proposed to workshop participants.

Speaker: James Yankel, Technical Director, Directorate of Logistics, Air Force Materiel Command

Presentation Title: Aging Aircraft Maintenance

The U.S. Air Force is going through a period of reduced recapitalization and increasing sustainment requirements as current fleets have lives extended 10 to 30 years into the future to maintain the current force structure. The efforts required to sustain these fleets is increasing both in material solution efforts, as failure modes beyond those identified in design begin to become more prevalent, and the resultant costs. Initiatives underway at Air Force Materiel Command (AFMC) are aggressively looking at sustainment of aging aircraft execution and issues. A discussion will be presented addressing the AFMC initiatives addressing organizational structure for sustainment, processes for sustainment as an enterprise, and technology development and insertion efforts for sustainment.

Speaker: Joann Berrett, Director, Aerospace Sustainment Directorate, Air Force Sustainment Center

Presentation Title: Driving to Cost-Effective Readiness

The Air Force Sustainment Center (AFSC) recognizes the challenges the U.S. Air Force continues to face to include aging aircraft fleets, the need to modernize weapon systems, rising weapon systems support costs, and fiscal constraints. We must reduce the cost of executing our mission. The cost of readiness will determine the size of our force, and the size of our force will determine our ability to fight and win the next war. AFSC is addressing the cost of executing its mission by focusing on cost-effective readiness through integrating efforts of organizations involved in sustainment activities. Recent success stories shared include reduction of flow days (or work-in-progress) for the KC-135 Stratotanker (the Air Force's primary in-flight refueling asset). This reduction in flow days resulted in the Oklahoma Air Logistics Center earning the 2011 Robert T. Mason Depot Excellence Award (first-ever Air Force winner). Another success story spotlighted a reduction in the number of C-130 Hercules "mission incapable" aircraft and improved due-date performance, another Robert T. Mason Depot Excellence Award Winner, but this time in 2012 (second Air Force winner in 2 years). Other success stories included improvements to the periodic depot maintenance for aircraft landing gear, concurrent E-3 Block 40/45 Modification installation (upgrade of electronics system on Airborne Warning and Control Systems), and propulsion alternate sourcing of parts. These efforts to date have provided seven additional KC-135s and 17 C-130s back to the field; reduced landing gear (Mission Incapable) MICAP hours by 91 percent; saved 8 months of depot possessed time for E-3s; and a cost avoidance of $65.4 million on propulsion parts.

Speaker: Mike Jennings, Deputy Director of Logistics (Acting), Air Force Life Cycle Management Center

Presentation Title: Product Support Responsibilities and Cost Reduction Initiatives

As the Air Force Life Cycle Management Center (AFLCMC) has evolved from planning efforts to standup, several initiatives have been pursued to ensure an emphasis on product support and the enterprise-wide role that now exists with the AFLCMC commander. As operations and support costs, including weapon system support costs, have continued to outpace inflation growth (FY1996-FY2011), AFLCMC now has an opportunity to execute an integrated, enterprise-wide product support strategy. Recommendations from the Life Cycle Management Working Group (LCM WG) and AFLCMC objective efforts can ensure enterprise visibility for the AFLCMC commander to exercise an influential role in product support activities across AFLCMC, as well as integrating product support management with the product support integrators and product support providers. AFLCMC has begun institutionalizing development of product support human capital, developmental planning, and logistics/health assessment recommendations from the LCM WG, as the AFLCMC commander has specific responsibilities in these areas. Other recommendations from the LCM WG will require a strengthening of AFLCMC commander roles/responsibilities. As initiatives progress, AFLCMC has the potential to ensure program office decisions are in-line with the Air Force enterprise product support strategy.

Speaker: Maj Mark Blumke, Deputy Chief, Mx Systems and Integration Branch, Directorate of Logistics, Air Mobility Command

Presentation Title: Air Mobility Command Initiatives to Reduce Maintenance and Sustainment Cost Drivers

No abstract submitted.

Speaker: SMSgt Kevin Mead, Air Force Element Vehicle and Equipment Management Support Office

Presentation Title: AFELM Vehicle and Equipment Management Support Office (VEMSO) Initiatives to Reduce Sustainment Costs

The Air Force is the fourth largest fleet within the federal government and supports more than 300 locations through 16 different regional headquarters. This results in significant duplication of effort and resources to meet the various missions. Knowing we cannot continue to operate this way in a fiscally challenging environment, Headquarters U.S. Air Force directed action to centralize management of the Air Force's vehicle fleet. The initiative is referred to as Installation Support Centralization (ISC). The ISC initiative consolidates enterprise vehicle fleet management and sustainment processes in support of logistics readiness operations, via Direct Liaison Authority (DIRLAUTH), with base level units, MAJCOM, and HAF staff. Consolidated activities are to be executed by AFELM VEMSO in the most efficient manner through continuous and deliberate reengineering of processes while enhancing global management and situational awareness of the Air Force's $7 billion vehicle fleet. Centralization has allowed AFELM VEMSO to implement several initiatives to reduce sustainment costs, including fleet validation/rightsizing and budget forecasting from an enterprise perspective. VEMSO is also leveraging technology to reduce costs with the implementation of the Automotive Information Management (AIM-2) system to collect vehicle data and automation of 10 fleet management processes to facilitate data integrity and reduce man-hour strains.

Speaker: Lt Col Brian Godfrey, Chief, Airborne Branch (A4CA), HQ Air Combat Command

Presentation Title: Air Combat Command Sustainment Challenges

Corrosion/structural cracks and wiring faults are factors driving sustainment costs in Air Combat Command (ACC). These factors are made worse by decreased manning, increased ops tempo, and aging fleets. Low observable (LO) maintenance in our fifth-generation fighter fleet continues to be an NM driver as well. ACC sees a potential for research into corrosion prevention, detection, treatment, and LO restoration to reduce sustainment costs. In addition, the ability to troubleshoot wiring and intermittent line replaceable unit faults with a test unit common across multiple airframes with more accuracy could greatly reduce sustainment costs. Finally, ACC would like to leverage state-of-the-art information technology tools, such as electronic tablets with mobile apps, to allow maintainers to reference tech data, conduct remote troubleshooting, order parts, close out work orders, etc., from the aircraft.

Speaker: BG Edward Dorman III, Director for Logistics Operations, Readiness, Force Integration and Strategy, Office of the Deputy Chief of Staff, G-4

Presentation Title: Informed Sustainment

The Army is developing "Lean" approaches to power and sustain its operations in the face of global, dynamic threats, while streamlining the force to meet resource constraints. Two examples, energy-informed operations and condition-based maintenance, tie investments and priorities to operational significance and risk. Energy enables all operational capabilities, but delivery incurs casualties and cost—about $2 billion per year in Southwest Asia. The Secretary of the Army has endorsed a new initiative to establish an "energy-informed" culture, which admonishes the total Army to behave in ways that maximize the net operational benefit from energy. This requires a combination of information, education, technologies, and processes to inform decisions and enable behaviors. Condition-based maintenance invokes a similar mission-informed decision process to focus reset efforts in order to manage risk and maintain a ready posture in the wake a decade of high operational tempo. Each of these initiatives illustrates a maturing Army capability to focus resources and priorities based upon mission demands.

Speaker: Joe Guenther, Vice President and General Manager, Evandale Turbofan and Turbojet Engines, General Electric Aviation

Presentation Title: GE Initiatives to Reduce Sustainment Costs

In 2011, the global fleet of military and commercial aircraft was powered by more than 50,000 engines provided by GE and its partners. GE Aviation is under a variety of service contracts to provide sustainment for, more than 11,000 of those commercial and military engines. Factors driving GE Aviation sustainment costs and how those factors change over time will be discussed. Examples of successful management of commercial and military engine sustainment costs will be examined, and emerging technologies that have demonstrated reduced sustainment costs will be presented. In closing, recommendations to take advantage of emerging capabilities will be provided.

Speaker: Raymond Valeika, Retired Senior Vice President for Technical Operations, Delta Airlines

Presentation Title: Sustainment: Managing the Consequences of Failure with Transparent Information

Sustainment or the maintenance of aircraft is a very complex business. The multiplicity of variables often drives actions and polices in a variety of directions. This then increases cost and often does not add to safety and availability of aircraft for the intended missions. This discussion deals with two fundamental issues in effective sustainment programs: understanding the consequence of failure and then assuring that proper information is available not only to understand the consequence of failure, but also to analyze life-cycle activities to determine their effectiveness. Consequence of failure determines the maintenance requirements. The analysis behind that is critical to ensure that both safety and economics are considered. Information is then the key for assuring the ongoing maintenance provides the necessary results. The free flow of information is critical in measuring and establishing goals and accountability. This paper discusses both the technical factors and organizational impacts of ongoing maintenance requirements and then gives some examples of programs that worked.

Speaker: Mike Buongiorno, Director, Military Engine Aftermarket Business Development, Pratt and Whitney (P&W)

Presentation Title: P&W Life Cycle Cost Management

With the continued emphasis on tight defense budgets both within the United States and our global allies, P&W remains aligned with our customers though our Integrated Program Deployment (IPD) strategy. Through IPD, P&W focuses on reducing the life-cycle cost of the propulsion system from product development, through production and into sustainment. P&W has demonstrated success in design for reliability/maintainability, production cost target achievement, and integrated sustainment solutions that not only reduce depot maintenance cost but also focus on reducing maintenance through increased time on wing, optimized operations and sustainment integration with original equipment manufacturer knowledge. P&W leverages lessons learned in legacy programs to drive continuous improvement into new products as well as flows new technology back into mature platforms to enhance their durability and reduce operating cost. P&W remains engaged with the services through the Component Improvement Program and other government and company-funded initiatives to reduce the cost of propulsion sustainment.

Speaker: VADM Walter Massenburg (USN, Ret.), Senior Director, Mission Assurance Business Education, Raytheon Company

Presentation Title: Enterprise, Why Now?: Naval Aviation Enterprise Model

In 1999, naval aviation was in crisis. As leaders of naval aviation in the 1990s prioritized building the future force structure to replace an aging aircraft fleet, the existing fleet continued to age, and the budget to preserve and manage the aging fleet was continually cut. Naval aviation faced the unprecedented crisis of having a force not ready to fight, while losing a generation of leadership. The "stovepipes" of operations, maintenance, and supply that contributed to current readiness retrenched and sought to optimize their activities at the expense of others. If not addressed, this "downward death spiral" would have resulted in a greatly reduced force structure, and warfighting capability would have been compromised. Across the years 1999-2007, naval aviation created a different business model which valued cost wise readiness and developed the concept of single process ownership and the single fleet driven metric to establish a horizontal behavior model that valued aviation units ready for tasking at reduced cost—today, tomorrow, and in the future. They adopted continuous process improvement (AIRSpeed), public private partnerships, performance based logistics, and other tools to enable the transformation. Today, the cost of naval aviation current readiness is predictable, billions of dollars remain in the Future Years Defense Program to recapitalize the force, and the level of readiness (availability) that is required is understood, achieved, and maintained. The Naval Aviation Enterprise operates as a true enterprise where readiness at reduced cost is everyone's responsibility. Today, the Air Force is experiencing the same crisis that naval aviation experienced in the late 1990S—aging aircraft fleets, austere budgets, and "stovepipes" that drive cost. This naval aviation enterprise concept is applicable in any government organization; but now, as service budgets face severe pressures, the Air Force could use this current crisis to adopt an enterprise model and change from a "business of consumption of resources" to one that values a "business of conservation of resources."